Lecture Notes in Economics and Mathematical Systems

548

Founding Editors:

M. Beckmann
H. P. Künzi

W0090835

Managing Editors:

Prof. Dr. G. Fandel
Fachbereich Wirtschaftswissenschaften
Fernuniversität Hagen
Feithstr. 140/AVZ II, 58084 Hagen, Germany

Prof. Dr. W. Trockel
Institut für Mathematische Wirtschaftsforschung (IMW)
Universität Bielefeld
Universitätsstr. 25, 33615 Bielefeld, Germany

Editorial Board:

A. Basile, A. Drexl, H. Dawid, K. Inderfurth, W. Kürsten, U. Schittko

Daniel Kuhn

Generalized Bounds for Convex Multistage Stochastic Programs

Springer

Author

Daniel Kuhn
Universität St. Gallen
Institut für Unternehmensforschung (HSG)
Bodanstraße 6
9000 St. Gallen
Switzerland

Library of Congress Control Number: 2004109705

ISSN 0075-8442
ISBN 3-540-22540-4 Springer Berlin Heidelberg New York

Springer is a part of Springer Science+Business Media

springeronline.com

Printed in Germany

Typesetting: Camera ready by author
Cover design: *Erich Kirchner*, Heidelberg

Printed on acid-free paper 42/3130Di 5 4 3 2 1 0

To Barbara

Preface

This work was completed during my tenure as a scientific assistant and doctoral student at the Institute for Operations Research at the University of St. Gallen. During that time, I was involved in several industry projects in the field of power management, on the occasion of which I was repeatedly confronted with complex decision problems under uncertainty. Although usually hard to solve, I quickly learned to appreciate the benefit of stochastic programming models and developed a strong interest in their theoretical properties. Motivated both by practical questions and theoretical concerns, I became particularly interested in the art of finding tight bounds on the optimal value of a given model. The present work attempts to make a contribution to this important branch of stochastic optimization theory. In particular, it aims at extending some classical bounding methods to broader problem classes of practical relevance.

This book was accepted as a doctoral thesis by the University of St. Gallen in June 2004. I am particularly indebted to Prof. Dr. Karl Frauendorfer for supervising my work. I am grateful for his kind support in many respects and the generous freedom I received to pursue my own ideas in research. My gratitude also goes to Prof. Dr. Georg Pflug, who agreed to co-chair the dissertation committee. With pleasure I express my appreciation for his encouragement and continuing interest in my work.

Of course, this book would not have achieved its present form without the support of my fellow colleagues. I enjoyed the numerous stimulating discussions with Karsten Linowsky about topics of scientific or secular interest. Moreover, I benefited a lot from the experience and expertise of Jens Güssow and Georg Ostermaier in the field of power management and applied stochastic optimization. A special thanks goes to Olivier Schmid and Patrick Wirth who shared with me their insights into the subtleties of the barycentric approximation scheme. I am also truly thankful to Manfred Grollmann for computer support and transportation services. Furthermore, I am indebted to

Michael Schürle who had always an open ear for my questions and ceaselessly regaled me with anecdotes from the past. A particular debt of thanks is owed to Gido Haarbrücker and Dominik Boos for continuously appealing to my mathematical conscience and to Ulrich Jacobi for competent assistance with computer problems. I also acknowledge the collegiality and manifold help of Massimo Cutaia and Jérôme Koller and the fruitful cooperation with Daniel Hofstetter during the doctoral seminars.

St. Gallen, June 2004 *Daniel Kuhn*

Contents

1 Introduction ... 1
1.1 Motivation ... 1
1.2 Previous Research ... 2
1.3 Objective ... 4
1.4 Outline ... 5

2 Basic Theory of Stochastic Optimization ... 7
2.1 Modelling Uncertainty ... 7
2.2 Policies ... 10
2.3 Constraints ... 11
2.4 Static and Dynamic Stochastic Programs ... 14
2.5 Here-and-Now Strategies ... 25
2.6 Wait-and-See Strategies ... 25
2.7 Mean-Value Strategies ... 26

3 Convex Stochastic Programs ... 29
3.1 Augmenting the Probability Space ... 29
3.2 Preliminary Definitions ... 33
3.2.1 Slater's Constraint Qualification ... 33
3.2.2 Convex Functions ... 35
3.2.3 Block-Diagonal Autoregressive Processes ... 36
3.3 Regularity Conditions ... 37
3.4 sup-Projections ... 40
3.5 Saddle Structure ... 41
3.6 Subdifferentiability ... 47

4 Barycentric Approximation Scheme ... 51
4.1 Scenario Generation ... 51
4.2 Approximation of Expectation Functionals ... 54
4.2.1 Jensen's Inequality ... 55
4.2.2 Edmundson-Madansky Inequality ... 57

4.2.3 Lower Barycentric Approximation 59
4.2.4 Upper Barycentric Approximation 62
4.3 Partitioning 63
4.4 Barycentric Scenario Trees 67
4.5 Bounds on the Optimal Value 74
4.6 Bounding Sets for the Optimal Decisions 77

5 Extensions 83
5.1 Stochasticity of the Profit Functions 84
5.2 Stochasticity of the Constraint Functions 89
5.3 Synthesis of Results 100
5.4 Linear Stochastic Programs 102
5.4.1 D.c. Functions 105
5.4.2 Generalized Bounds for Linear Stochastic Programs 106
5.5 Bounding Sets for the Optimal Decisions 111

6 Applications in the Power Industry 113
6.1 The Basic Decision Problem of a Hydropower Producer 115
6.2 Market Power 118
6.3 Lognormal Spot Prices 120
6.4 Lognormal Natural Inflows 121
6.5 Risk Aversion 123
6.6 Numerical Results 126
6.6.1 Model Parameterization 126
6.6.2 Discretization of the Probability Space 128
6.6.3 Results of the Reference Problem 129
6.6.4 Hydro Scheduling Problem with Market Power 130
6.6.5 Lognormal Prices 131
6.6.6 Hydro Scheduling Problem with Lognormal Inflows 133
6.6.7 Hydro Scheduling Problem with Risk-Aversion 137

7 Conclusions 141
7.1 Summary of Main Results 141
7.2 Future Research 145

A Conjugate Duality 147

B Lagrangian Duality 155

C Penalty-Based Optimization 163

D Parametric Families of Linear Functions 165

E Lipschitz Continuity of sup-Projections 169

References 175

List of Figures 183

List of Tables 185

Index 187

1

Introduction

1.1 Motivation

Almost any technical or economic decision problem includes some degree of uncertainty about the values to assign to some problem-specific parameters. The best decision strategies with respect to some objective criterion must be found on the basis of the a priori information about these uncertainties. If it is possible to assign a probability distribution to the random parameters, the determination of an optimal decision strategy gives rise to a stochastic optimization model, also referred to as a *stochastic program*. However, the solution of stochastic programs poses severe difficulties, especially in the multistage case. If the underlying probability space is continuous, the stochastic program represents an optimization problem over an infinite-dimensional function space. Then, analytical solutions are hardly available, and nontrivial problems of practical relevance must always be solved numerically. However, numerical solution requires discretization of the continuous probability space. One should select a discrete probability measure with finite support and solve the stochastic program with respect to this discrete auxiliary measure, instead of the continuous original measure. In doing so, one effectively approximates the original stochastic program by an optimization problem over a finite-dimensional Euclidean space, which is numerically tractable.

The selection of an appropriate discrete probability measure is referred to as *scenario generation* and represents a primary challenge in the field of stochastic programming. It is indispensable that the solution of the approximate problem can be related in some way to the solution of the original problem, i.e. the exact solution of the auxiliary problem should provide an approximate solution of the original stochastic program. Ideally, one can find a discrete probability measure such that the optimal value and the optimizer set of the associated auxiliary stochastic program are, in a quantitative sense, close to the optimal value and the optimizer set of the original optimization problem, respectively. However, there is always a tradeoff between accuracy, requiring as many discretization points as possible, and computational effort, which increases dramatically with the number of discretization points.

In this work we will always assume that the stochastic program under consideration has several decision stages. For the sake of a suggestive terminology, we will moreover assume that the objective criterion is to maximize expected profit. At each decision stage a realization of some stochastic parameter is observed in response to which a decision is selected. Typically, the decision is chosen so as to maximize the sum of a certain immediate profit and the conditional expectation of an uncertain future profit. In addition, the decision is subject to specific constraints, which may depend on the random parameters observed by that time.

The recourse function (profit-to-go function) of a specific stage is defined as the expected future profit conditional on the observations up to that stage and the decisions selected in the previous stages. As in dynamic programming, the recourse functions can be calculated recursively by iterative application of the operations 'maximization' and 'integration' with respect to some conditional probability measure. Knowledge of the structural properties of the recourse functions (measurability, convexity, concavity, subdifferentiability, Lipschitz continuity, etc.) is of fundamental importance for efficient scenario generation.

1.2 Previous Research

Let us briefly summarize some scenario generation methods which have received considerable attention in the stochastic programming literature. We primarily focus on bounding methods and optimal discretization by using probability metrics. Other approaches (e.g. conditional sampling, moment matching, path-based methods, etc.) are of minor importance for our purposes and will therefore be omitted in the subsequent discussion. A survey of these methods is provided in [62].

First, we address the class of bounding methods, which enjoy growing popularity since the 1960s. Thereby, one attempts to construct two discrete probability measures such that the optimal values of the associated approximate problems represent upper and lower bounds for the optimal value of the original stochastic program, respectively. In developing bounding probability measures, one typically exploits structural properties of the recourse functions. When the recourse functions of all decision stages are convex in the random parameters, a lower bounding measure can be found by means of Jensen's inequality [57], which is based on utilizing only first moment information. Notice that Jensen's inequality applies under fairly general conditions, e.g. for multivariate random vectors with correlated components and an unbounded support. If, however, the support of the random variables represents a compact parallelepiped, a first order upper bound is provided via the Edmundson-Madansky inequality; Edmundson [35] concentrates on the univariate case, whereas Madansky [69,70] and Frauendorfer [41] consider the multivariate setting, given that the components of the random vector are independent and dependent, respectively. In the dependent case, not only the mean but also the first order cross moments of the underlying random vector are used. Gassmann and Ziemba [47] generalize the Edmundson-Madansky bound to dependent random variables defined on polyhedral sets, while Birge and Wets [11,12] allow for unbounded domains.

When the difference between upper and lower bounds is larger than some prescribed tolerance, one can try to incorporate higher order information in the construction of bounding measures. Dokov and Morton [23] develop lower bounds based on second order information, while Birge and Dulá [9], Dupačová [30], and Kall [59] propose second order upper bounds. Higher order upper bounds are provided in [24]. All of the bounds mentioned so far can be improved to an arbitrary degree of precision by applying them on sufficiently small subsets of the underlying domain. This technique – known as *partitioning* – is exemplified in [11,45,56].

If the recourse functions are concave, Jensen's inequality and its generalizations yield upper bounds, whereas the Edmundson-Madansky-type inequalities furnish lower bounds. This is a mere consequence of the fact that multiplication by -1 maps any convex function to a concave function and vice versa.

The above bounds can be generalized to the case when the recourse functions are convex-concave saddle functions. First order bounds similar to those of Jensen and Edmundson-Madansky are developed by Frauendorfer [42] under the assumption that the underlying random vector is defined on a multidimensional rectangle or a cross-simplex (i.e. the Cartesian product of two simplices). Edirisinghe and Ziemba [33, 34] generalize these bounds to random variables on an arbitrary polyhedral probability space. Moreover, Edirisinghe [31] constructs bounding measures based on second order information; see also the similar approach presented in [111]. Suitable partitioning schemes for bounds on saddle functions are proposed in [32,42].

Notice that several of the bounding measures presented above may be derived as solutions of generalized moment problems [12,33,34,42,58]. This interpretation of bounds for stochastic programs originates from the work of Dupačová in a game-theoretic setting [30] and also has applications in models where the underlying probability distribution is only known in limited manner [25–27].

The classical bounding methods for scenario generation fail unless the recourse functions of a given stochastic program can be shown to be convex, concave, or saddle-shaped in the stochastic parameters. In order to solve more general optimization problems, one should relax this sensible requirement. For instance, if the recourse functions exhibit some local Lipschitzian properties, one can apply scenario generation by optimal discretization due to Pflug [85]. This approach aims at finding a discrete probability measure with a prescribed number of mass points which is close to the original measure with respect to some probability metric. Then, one can prove that the optimal value of the original problem differs from the optimal value of the discrete auxiliary problem at most by the distance of the two measures, with respect to the given probability metric, multiplied by an appropriate Lipschitz constant. Rachev and Römisch [87] use a similar method to optimally reduce the number of mass points of a discrete probability measure obtained via sampling; see also [29]. This scenario reduction technique can occasionally decrease computational complexity of a stochastic program without seriously affecting its performance.

1.3 Objective

From the point of view of applications, most of the scenario generation methods presented in the previous section suffer from more or less serious deficiencies. For example, use of the classical bounding methods is restricted to problems with either convex, concave, or – more generally – saddle-shaped recourse functions. Loosely speaking, the following standard assumptions suffice to imply the saddle property of the recourse functions (see Chap. 3):

(a) the stochastic program represents a convex optimization problem (without loss of generality we focus on maximization problems);

(b) the immediate profit in each stage represents a convex function of the stochastic parameters;

(c) the constraint functions are jointly convex in the decision variables and the stochastic parameters;

(d) the stochastic parameters follow autoregressive processes driven by serially uncorrelated noise.

Although the first requirement seems to be fairly restrictive, many important real-life decision problems can be formulated as convex stochastic programs. Conversely, the conditions (b)–(d) are restrictive, especially in the multistage case. For example, stochastic programs involving lognormally distributed prices and demands, derivative trading, or risk-aversion do not meet the requirements (b), (c), and (d) jointly. While being practically relevant, such problems go beyond the scope of the classical bounding methods.

Notice that the modern methods based on probability metrics apply to broader classes of decision problems. For a brief outline of the limitations of these approaches we refer to [62, Sect. 4].

The present work concentrates on bounding methods. In particular, we will use the *barycentric approximation scheme* by Frauendorfer [42–44] as a starting point. This method is adapted for problems which satisfy the conditions (a)–(d) and provides two complementary discrete probability measures. Solution of a stochastic program with respect to these barycentric measures yields lower and upper bounds for the true optimal value, respectively. By successively partitioning the domain of the random vectors, the bounds can be made arbitrarily tight.

In this book we interpret the barycentric discretizations and the associated auxiliary stochastic programs in a more general setting. Concretely speaking, we drop the restrictive assumptions (b) and (c), thus investigating convex multistage stochastic programs with a generalized nonconvex dependence on the random variables. Then, the optima of the discretized auxiliary problems can no longer be shown to represent strict bounds on the true optimal value. However, we will argue that they still provide bounds after a simple transformation. The main goal of the present work is to characterize this transformation in an intuitive though mathematically rigorous way. In this regard, we will formulate weak regularity conditions under which the new bounding method is applicable. Thereby, we identify broad problem classes for which easily computable bounds are available. Moreover, it will be shown that the new bounds can still be made tight via partitioning.

As a second major objective, we present a collection of important real-life decision problems in the field of power management that can be addressed with our new approach. These problems go beyond the scope of the classical bounding methods. Their formulation as multistage stochastic programs requires specific modelling techniques, which will be investigated systematically. Numerical calculations are provided for illustrative purposes.

1.4 Outline

This book is structured as follows. In Chap. 2 we briefly review the basic theory of stochastic optimization. After some introductory remarks about the mathematical representation of uncertainty, we give precise definitions of decision strategies, constraints, and objective functions. Subsequently, the static and dynamic versions of a general non-linear stochastic program are formulated, and a set of fundamental regularity conditions is introduced. Under these regularity conditions, the static and dynamic versions of a stochastic program can be shown to be well-defined, solvable, and equivalent. At the end of the chapter we will sketch a few approaches to evaluate the benefit of using a stochastic optimization model.

Chapter 3 specializes the basic theory of stochastic optimization to the important class of *convex* stochastic programs. Here, we work with a set of stronger regularity conditions which provide sharper results than in Chap. 2. Given these stronger conditions, the recourse functions of any convex stochastic program are subdifferentiable and exhibit a characteristic saddle structure. The presented theorems slightly generalize some well-known results.

Chapter 4 deals with scenario generation based on classical barycentric approximation. We start by studying generalized moment problems whose dual solutions represent extremal probability measures. These extremal measures can be used to synthesize two discrete bounding measures for stochastic programs which satisfy the regularity conditions of Chap. 3. Next, we recover the main result of [43], i.e. the auxiliary stochastic programs associated with the discrete barycentric measures provide bounds on the true optimal value. Moreover, we propose a slightly generalized partitioning method to improve the bounds. In conclusion, we introduce a sequence of bounding sets for the optimal first-stage decisions. This sequence is shown to converge to the maximizer set of the underlying stochastic program. Our analysis sharpens the epi-convergence results of Birge and Wets [11].

In Chap. 5 we relax the regularity conditions of Chap. 3 to allow for convex stochastic programs with a nonconvex dependence on the stochastic parameters. Under weak assumptions, such 'irregular' problems can be transformed to 'regular' problems, whose recourse functions are subdifferentiable and exhibit a characteristic saddle structure. The classical barycentric bounds for the regular problems can then be back-transformed to yield bounds on the corresponding irregular problems. In the case of linear stochastic programs, we derive a particularly intuitive formula for the new bounds.

Chapter 6 presents exemplary real-life applications of the theoretical concepts developed in Chap. 5. Concretely speaking, it will be shown how market power, lognormal stochastic processes, and risk-aversion can be properly handled in a stochastic programming framework. After a detailed exposition of appropriate modelling techniques, we report on numerical experience with the new bounding method. Finally, Chap. 7 concludes.

2

Basic Theory of Stochastic Optimization

In this chapter we develop the basic theory of stochastic optimization. After some introductory remarks about the mathematical representation of uncertainty, we investigate the key ingredients of general stochastic programs, i.e. decision strategies, constraints, and objective functions. Subsequently, the static and dynamic versions of a stochastic optimization problem are formulated, and some elementary regularity conditions are discussed. Under these regularity conditions, the static and dynamic versions of the stochastic program at hand can be shown to be well-defined, solvable, and equivalent. Finally, we discuss two useful indicators, which enjoy wide popularity in literature: the *expected value of perfect information* (EVPI) represents the maximum amount to be paid in return for complete and accurate information about the future, whereas the *value of the stochastic solution* (VSS) quantifies the cost of ignoring uncertainty in choosing a decision.

2.1 Modelling Uncertainty

Discrete-time multistage stochastic programs are built on a probability space $(\hat{\Omega}, \hat{\mathcal{F}}, \hat{P})$ equipped with a *filtration*, i.e. an increasing sequence of σ-algebras $\{\hat{\mathcal{F}}^t\}_{t\in\tau}$ included in $\hat{\mathcal{F}}$. Thus, we have $\hat{\mathcal{F}}^s \subseteq \hat{\mathcal{F}}^t$ for all $s \leq t$ lying within the finite index set $\tau := \{0, \ldots, T\}$. Sometimes, we will also need the related index sets $\tau_{-t} := \tau\backslash\{t\}$, $t \in \tau$. The parameter t is normally interpreted as a time index, and T usually indices the last decision stage of a given stochastic program. Without loss of generality we may postulate $\hat{\mathcal{F}}^T = \hat{\mathcal{F}}$. In case of a finite probability space, it is usually assumed that the σ-algebra $\hat{\mathcal{F}}$ coincides with the power set $2^{\hat{\Omega}}$. The measurable space $(\hat{\Omega}, \hat{\mathcal{F}})$ is called *sample space*; elements of $\hat{\Omega}$ are called *outcomes*, and elements of $\hat{\mathcal{F}}$ are referred to as *events*. Furthermore, the σ-algebra $\hat{\mathcal{F}}^t$ can conveniently be seen as the information available at time t or as the *σ-algebra of events* up to time t.

A *random variable* or a *random vector* is defined as a measurable function $\tilde{\omega} : (\hat{\Omega}, \hat{\mathcal{F}}) \to (\Omega, B(\Omega))$, where Ω denotes the *state space* comprising the so-called *observations*. In the context of stochastic programming, Ω is usually taken to be a compact subset of $\mathbb{R}^M$, and $\mathcal{B}(\Omega)$ denotes the Borel field[1] of the state space. Furthermore, a *stochastic process* is understood to be a family of random variables $\{\tilde{\omega}_t\}_{t\in\mathcal{T}}$, $\tilde{\omega}_t : (\hat{\Omega}, \hat{\mathcal{F}}) \to (\Omega_t, \mathcal{B}(\Omega_t))$. Again, Ω_t is supposed to be a compact subset of $\mathbb{R}^{M_t}$. By convention, a stochastic process is called $\hat{\mathcal{F}}^t$*-adapted* or $\hat{\mathcal{F}}^t$*-previsible* if $\tilde{\omega}_t$ is measurable with respect to $\hat{\mathcal{F}}^t$ or $\hat{\mathcal{F}}^{t-1}$, respectively. In many applications, $\tilde{\omega}_0$ is measurable with respect to the trivial σ-algebra $\{\emptyset, \hat{\Omega}\}$. This implies roughly that there is no information at time 0. By definition, a sequence of $\{\emptyset, \hat{\Omega}\}$-measurable random variables represents a *deterministic* process.

A filtration $\{\hat{\mathcal{F}}^t\}_{t\in\mathcal{T}}$ is said to be *induced* by a process $\{\tilde{\omega}_t\}_{t\in\mathcal{T}}$ if $\hat{\mathcal{F}}^t$ coincides with the σ-field generated by the sets $\cup_{s=0}^t\{\tilde{\omega}_s^{-1}(A)|A \in \mathcal{B}(\Omega_s)\}$. The induced σ-algebra $\hat{\mathcal{F}}^t$ describes the information which is available at time t by only observing the underlying process. Figure 2.1 shows an exemplary discrete stochastic process $\{\tilde{\omega}_t\}_{t\in\mathcal{T}}$ with four time steps. Every different path of the corresponding scenario tree, i.e. every possible sequence of observations, is assigned to one element of the sample space, which is chosen to be $\hat{\Omega} = \{\hat{\omega}_1, \ldots, \hat{\omega}_8\}$. The atoms of the induced σ-field $\hat{\mathcal{F}}^t$ can be determined by forming the union of all paths which are indistinguishable up to time t. In Fig. 2.2, the samples belonging to an atom of $\hat{\mathcal{F}}^t$ are linked with a shaded line. By construction, all atoms of stage $t+1$ are subsets of the atoms of stage t, implying that $\hat{\mathcal{F}}^t \subseteq \hat{\mathcal{F}}^{t+1}$. This sequential inclusion reflects the idea that events are never 'forgotten', i.e. a filtration basically determines how information is revealed through time.

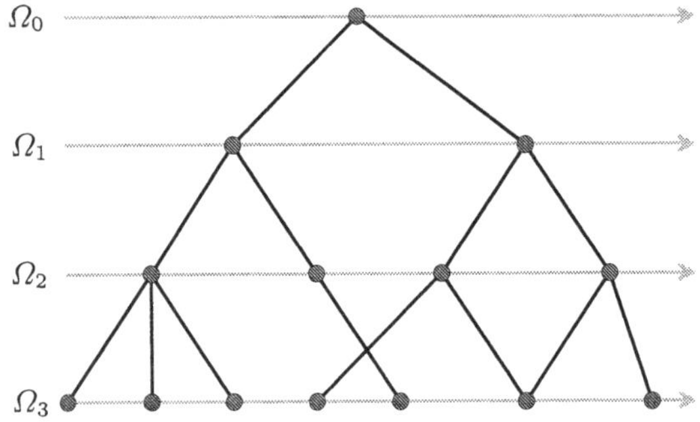

Fig. 2.1. Example of a stochastic process $\{\tilde{\omega}_t\}_{t\in\mathcal{T}}$

[1]The Borel field $\mathcal{B}(\Omega)$ is the smallest σ-algebra containing all relatively open subsets of Ω with respect to the standard topology of $\mathbb{R}^M$.

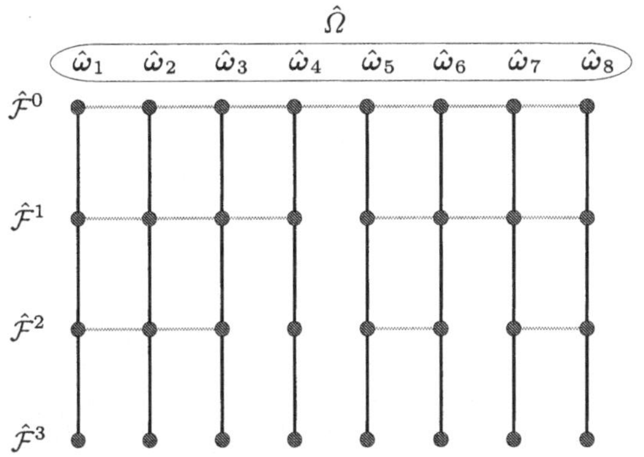

Fig. 2.2. Visualization of the filtration induced by $\{\tilde{\omega}_t\}_{t\in\tau}$

For later use we define a random variable $\tilde{\omega}$ corresponding to the stochastic process $\{\tilde{\omega}_t\}_{t\in\tau}$.

$$\tilde{\omega} := \tilde{\omega}_0 \times \cdots \times \tilde{\omega}_T : \hat{\Omega} \to \Omega_0 \times \cdots \times \Omega_T =: \Omega \tag{2.1}$$

Every element of the state space Ω can be identified with an equivalence class of outcomes in the sample space. Thus, $(\Omega, \mathcal{F})$ naturally inherits the probability measure defined on $(\hat{\Omega}, \hat{\mathcal{F}})$ through $P(A) := \hat{P}(\tilde{\omega}^{-1}(A))$, $A \in \mathcal{F} := \mathcal{B}(\Omega)$. By construction, P is a regular probability measure on $(\Omega, \mathcal{F})$. Moreover, the state space is equipped with a filtration $\{\mathcal{F}^t\}_{t\in\tau}$, which is given through

$$\mathcal{F}^t := \left\{ A \times \Omega_{t+1} \times \cdots \times \Omega_T \,\middle|\, A \in \mathcal{B}(\Omega_0 \times \cdots \times \Omega_t) \right\}.$$

Instead of the abstract probability space $(\hat{\Omega}, \hat{\mathcal{F}}, \hat{P})$, we may equivalently consider the induced probability space $(\Omega, \mathcal{F}, P)$, which we will use henceforth. With this convention, $\tilde{\omega}$ reduces to the identity map, while $\tilde{\omega}_t$ becomes a specific coordinate projection for every $t \in \tau$. Moreover, the terms 'outcome' and 'observation' will from now on be used synonymously. From a conceptual point of view, it is important to distinguish random variables $\tilde{\omega}_t$ and their realizations, which will be denoted by ω_t below.

By convention, $E(\cdot)$ denotes expectation over the probability measure P. Conditional expectations $E_t(\cdot) := E(\cdot|\mathcal{F}^t)$ on $(\Omega, \mathcal{F}, P)$ are defined up to an equivalence relation; i.e. there can be many versions of $E_t(\cdot)$, which differ on P-null sets. In this work, $E_t(\cdot)$ is taken to be a *regular* conditional expectation being representable as an indefinite integral with respect to a *regular* conditional probability. Such regular conditional probabilities exist since $\mathcal{F}$ is the Borel field on Ω and P is a regular Borel measure [66, Sect. 27].

2.2 Policies

In the sequel we assume $\{\tilde{\omega}_t\}_{t\in\mathcal{T}}$ to characterize some uncertain problem data with respect to future time periods. In the context of power management, for instance, electricity prices, load demand, reservoir inflows, and fuel prices are major uncertain impacts contained in $\tilde{\omega}_t$. Information on the random data becomes available successively at finitely many time points, at which decisions are selected. After a first observation $\boldsymbol{\omega}_0$, an initial decision $\boldsymbol{x}_0$ is taken. At this stage, the decision maker has no information about future outcomes. Then, a second observation $\boldsymbol{\omega}_1$ is made, in response to which a subsequent decision is selected, etc. Generally speaking, after the observation of the outcomes $(\boldsymbol{\omega}_0,\ldots,\boldsymbol{\omega}_t)$, the decision maker selects some actions $\boldsymbol{x}_t \in \mathbb{R}^{n_t}$ according to a specific *decision rule.*[2] Such a decision rule depends on the nature of the underlying problem. Normally, rational decision makers choose actions maximizing some objective function. In the remainder of this section we aim at formalizing the concept of decision rules in a slightly more general setting, allowing also for so-called *anticipative* policies. By definition, an anticipative decision rule is a sequence of essentially bounded Borel measurable functions $\{\tilde{\boldsymbol{x}}_t\}_{t\in\mathcal{T}}$, i.e.

$$\tilde{\boldsymbol{x}}_t \in \mathcal{L}^\infty_{n_t} := \mathcal{L}^\infty(\Omega,\mathcal{F},P;\mathbb{R}^{n_t}). \tag{2.2}$$

Obviously, such a decision rule assigns a well-defined action vector to every time stage and every possible outcome. With the definitions of Sect. 2.1, any decision rule can be interpreted as a stochastic process[3] $\{\tilde{\boldsymbol{x}}_t\}_{t\in\mathcal{T}}$ on $(\Omega,\mathcal{F},P)$. By convention, $\boldsymbol{x}_t \in \mathbb{R}^{n_t}$ denotes a realization of $\tilde{\boldsymbol{x}}_t$. In order to simplify notation, we introduce the combined random variable $\tilde{\boldsymbol{\omega}}^t := (\tilde{\boldsymbol{\omega}}_0,\ldots,\tilde{\boldsymbol{\omega}}_t)$; its realizations $\boldsymbol{\omega}^t := (\boldsymbol{\omega}_0,\ldots,\boldsymbol{\omega}_t) \in \Omega^t := \times_{s=0}^t \Omega_t$ describe the sequence of observations or the *outcome history* up to time t ($\boldsymbol{\omega}^t \in \Omega^t \subset \mathbb{R}^{M^t}$, where $M^t := M_0 + \cdots + M_t$). Notice that $\tilde{\boldsymbol{\omega}}^T$ coincides with the random vector $\tilde{\boldsymbol{\omega}}$ defined in (2.1); a similar identity holds for the underlying domain, i.e. we have $\Omega^T = \Omega$, which is embedded in a Euclidean space of dimension $M := M^T$. Next, we adopt an analogous notation for the actions. By definition, $\tilde{\boldsymbol{x}}^t := (\tilde{\boldsymbol{x}}_0,\ldots,\tilde{\boldsymbol{x}}_t)$ represents a collection of measurable functions and $\boldsymbol{x}^t := (\boldsymbol{x}_0,\ldots,\boldsymbol{x}_t)$ denotes the *decision history* up to time t ($\boldsymbol{x}^t \in \mathbb{R}^{n^t}$, where $n^t := n_0 + \cdots + n_t$). In particular, we introduce the policy function $\tilde{\boldsymbol{x}} := \tilde{\boldsymbol{x}}^T$, which constitutes an n-dimensional Borel measurable mapping ($n := n^T$) and completely determines the underlying decision rule.

Let us now characterize *non-anticipative* decision rules. The natural causality structure described above, i.e. the requirement that future events

[2]A decision rule is also referred to as a *policy* or a *strategy.*

[3]Notice that the composed mapping $\tilde{\boldsymbol{x}}_t \circ \tilde{\boldsymbol{\omega}} : \hat{\Omega} \to \mathbb{R}^{n_t}$ is $\hat{\mathcal{F}}$-measurable, since $\tilde{\boldsymbol{x}}_t$ is assumed to be Borel measurable. This implies that $\tilde{\boldsymbol{x}}_t \circ \tilde{\boldsymbol{\omega}}$ is a random variable on the sample space $\hat{\Omega}$. Remember, in contrast, that compositions of Lebesgue measurable functions are generally not Lebesgue measurable.

may not influence present decisions, leads to a special functional dependence of the policy function upon the uncertain parameters. Formally speaking, $\tilde{\boldsymbol{x}}_t$ must be constant as a function of $\boldsymbol{\omega}_{t+1}, \ldots, \boldsymbol{\omega}_T$ for every $t \in \tau$. The space of non-anticipative decision rules is thus defined as (see e.g. [96])

$$\mathcal{N}_n := \mathcal{N}_{n_0} \times \mathcal{N}_{n_1} \times \cdots \times \mathcal{N}_{n_T},$$

where

$$\mathcal{N}_{n_t} := \mathcal{L}^\infty(\Omega, \mathcal{F}^t, P; \mathbb{R}^{n_t}) \quad \forall t \in \tau.$$

Obviously, $\mathcal{N}_n$ is a linear subspace of $\mathcal{L}_n^\infty = \mathcal{L}_{n_0}^\infty \times \cdots \times \mathcal{L}_{n_T}^\infty$. With a slight abuse of notation,[4] non-anticipative policies can be represented as

$$\tilde{\boldsymbol{x}}(\boldsymbol{\omega}) = \left(\tilde{\boldsymbol{x}}_0(\boldsymbol{\omega}^0), \tilde{\boldsymbol{x}}_1(\boldsymbol{\omega}^1), \ldots, \tilde{\boldsymbol{x}}_t(\boldsymbol{\omega}^t), \ldots, \tilde{\boldsymbol{x}}_T(\boldsymbol{\omega}^T)\right). \tag{2.3}$$

Notice that any non-anticipative decision process $\{\tilde{\boldsymbol{x}}_t\}_{t\in\tau}$ is adapted to the filtration induced by the random variables $\{\tilde{\boldsymbol{\omega}}_t\}_{t\in\tau}$. Thus, the predicates 'non-anticipative' and '$\mathcal{F}^t$-adapted' are equivalent characterizations of decision processes.

In a dynamic setting, the decisions $\boldsymbol{x}_t$ met at time t not only depend on the past outcomes $\boldsymbol{\omega}^t$ but also on the earlier decisions $\boldsymbol{x}^{t-1}$. In turn, these previous decisions depend on $(\boldsymbol{\omega}^{t-1}, \boldsymbol{x}^{t-2})$. Tracking the interdependence of decisions and outcomes inductively until the first stage, one can easily verify that (2.3) correctly reflects the net-dependence on the stochastic parameters.

2.3 Constraints

In broad problem classes the admissible actions at time t are subject to restrictions, which may depend on earlier decisions and on observations up to time t. Such restrictions are conveniently captured via vector-valued *constraint functions*. To each decision stage $t \in \tau$ we assign three Borel measurable constraint functions

$$\begin{aligned}
\boldsymbol{f}_t^{\text{in}} &: \mathbb{R}^{n^t} \times \mathbb{R}^{M^t} \to \mathbb{R}^{r_t^{\text{in}}}, \\
\boldsymbol{f}_t^{\text{eq}} &: \mathbb{R}^{n^t} \times \mathbb{R}^{M^t} \to \mathbb{R}^{r_t^{\text{eq}}}, \\
\boldsymbol{f}_t &: \mathbb{R}^{n^t} \times \mathbb{R}^{M^t} \to \mathbb{R}^{r_t}.
\end{aligned}$$

Thereby, it is assumed that $\boldsymbol{f}_t^{\text{in}}$ characterizes 'pure' inequality constraints, whereas $\boldsymbol{f}_t^{\text{eq}}$ determines possible equality constraints. By definition, the mapping $\boldsymbol{f}_t := (\boldsymbol{f}_t^{\text{in}}, \boldsymbol{f}_t^{\text{eq}}, -\boldsymbol{f}_t^{\text{eq}})$ accounts for both types of constraints. Then,

[4] There is a natural injection from the essentially bounded functions on Ω^t to the essentially bounded functions on Ω.

consistency requires the dimensions to match up properly, i.e. $r_t := r_t^{\text{in}} + 2\, r_t^{\text{eq}}$ for all $t \in \tau$. The sets of feasible first stage decisions are given through

$$X_0(\boldsymbol{\omega}_0) := \{\boldsymbol{x}_0 \in \mathbb{R}^{n_0} \,|\, \boldsymbol{f}_0(\boldsymbol{x}_0, \boldsymbol{\omega}_0) \leq \boldsymbol{0}\}, \tag{2.4a}$$

whereas the feasible sets of the recourse decisions are defined as

$$X_t(\boldsymbol{x}^{t-1}, \boldsymbol{\omega}^t) := \{\boldsymbol{x}_t \in \mathbb{R}^{n_t} \,|\, \boldsymbol{f}_t(\boldsymbol{x}^t, \boldsymbol{\omega}^t) \leq \boldsymbol{0}\} \quad \text{for } t \in \tau_{-0}. \tag{2.4b}$$

The above specifications allow for a uniform treatment of equality and inequality constraints, since any equation of the form $\boldsymbol{f}_t^{\text{eq}} = \boldsymbol{0}$ is equivalent to two opposing inequalities $\boldsymbol{f}_t^{\text{eq}} \leq \boldsymbol{0}$ and $-\boldsymbol{f}_t^{\text{eq}} \leq \boldsymbol{0}$. Notice that the *feasible set mappings*

$$\boldsymbol{\omega}_0 \mapsto X_0(\boldsymbol{\omega}_0) \quad \text{and} \quad (\boldsymbol{x}^{t-1}, \boldsymbol{\omega}^t) \mapsto X_t(\boldsymbol{x}^{t-1}, \boldsymbol{\omega}^t) \quad \text{for } t \in \tau_{-0}$$

can be viewed as set-valued functions or multifunctions. For any $t \in \tau$ the *nested feasible set mapping* X^t is defined through

$$X^t : \begin{cases} \Omega^t \to 2^{\mathbb{R}^{n^t}} \\ \boldsymbol{\omega}^t \mapsto \{\boldsymbol{x}^t \in \mathbb{R}^{n^t} \,|\, \boldsymbol{x}_0 \in X_0(\boldsymbol{\omega}_0), \ldots, \boldsymbol{x}_t \in X_t(\boldsymbol{x}^{t-1}, \boldsymbol{\omega}^t)\}. \end{cases}$$

By construction, the values of X^t are closed subsets of $\mathbb{R}^{n^t}$ if the constraint functions are lower semicontinuous in the decision variables. Moreover, X^t constitutes an $\mathcal{F}^t$-measurable multifunction if the constraint functions are *normal integrands* in the sense of [91, theorem 2J]. The set-valued mapping $X := X^T$ is referred to as *constraint multifunction* in literature, see e.g. Rockafellar and Wets [92]. In case of a well-defined decision problem, the multifunction X_t is non-empty-valued for any thinkable outcome and decision history ($t \in \tau$). Formally speaking, for any possible sequence of outcomes $\boldsymbol{\omega}^t \in \Omega^t$ and for any admissible decision history $\boldsymbol{x}^{t-1} \in X^{t-1}(\boldsymbol{\omega}^{t-1})$ we require $X_t(\boldsymbol{x}^{t-1}, \boldsymbol{\omega}^t) \neq \emptyset$. As easily can be verified, this characterization of well-definedness is equivalent to the condition

$$\{\boldsymbol{x}^t \in \mathbb{R}^{n^t} \,|\, \boldsymbol{x} \in X(\boldsymbol{\omega})\} \equiv X^t(\boldsymbol{\omega}^t) \quad \forall \boldsymbol{\omega} \in \Omega, t \in \tau. \tag{2.5}$$

Equation (2.5) implies that the projection of $X(\boldsymbol{\omega})$ on the action space of the first t stages only depends on $\boldsymbol{\omega}^t$. Therefore, a constraint multifunction which satisfies condition (2.5) is called *non-anticipative* [92]. In particular, this condition implies X to be non-empty-valued on the entire probability space Ω if the first stage feasible set $X_0(\boldsymbol{\omega}_0)$ is non-empty for all $\boldsymbol{\omega}_0 \in \Omega_0$. If (2.5) fails to be true, all decisions that lead to infeasibilities in certain scenarios of future time stages must a priori be excluded by so-called *induced constraints.* Notice that every constraint multifunction can be made non-anticipative by introducing explicitly the induced constraints, as pointed out in [94, 107]. In this work we shall confine ourselves to non-anticipative decision problems since induced constraints are acausal and should be avoided in realistic models.

One might be tempted to relax condition (2.5) to hold only on the support of the probability measure P instead of the entire space Ω (note that $\operatorname{supp} P$ can be a strict subset of Ω). However, the scenario generation method to be developed in Sect. 4 relies on the assumption that the constraint multifunction satisfies condition (2.5) for specific scenarios in $\Omega \backslash \operatorname{supp} P$. As a matter of fact, for most problems of practical interest this extra requirement is no real restriction.

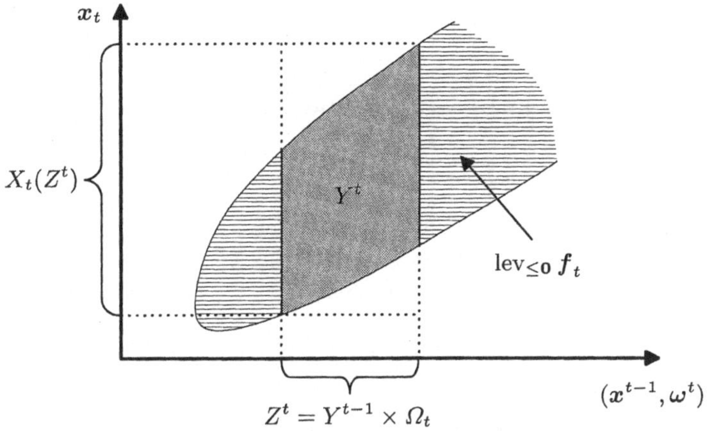

Fig. 2.3. Y^t is the graph of the multifunction X_t over Z^t and lies in the level set $\text{lev}_{\leq 0}\, \boldsymbol{f}_t$. The image of Z^t under X_t is given by the projection of Y^t on $\mathbb{R}^{n_t}$

When dealing with constrained multistage stochastic programs, it is sometimes useful to introduce a sequence of *generalized feasible sets* Y^t comprising the joint outcome and decision histories feasible up to time t.

$$Y^t := \left\{ (\boldsymbol{x}^t, \boldsymbol{\omega}^t) \,\middle|\, \boldsymbol{\omega}^t \in \Omega^t,\, \boldsymbol{x}^t \in X^t(\boldsymbol{\omega}^t) \right\} \quad \forall t \in \tau \tag{2.6}$$

For the sake of transparent notation, we need the related *generalized feasible sets* Z^t containing no stage t decisions. Define $Z^0 := \Omega_0$ and

$$Z^t := \left\{ (\boldsymbol{x}^{t-1}, \boldsymbol{\omega}^t) \,\middle|\, \boldsymbol{\omega}^t \in \Omega^t,\, \boldsymbol{x}^{t-1} \in X^{t-1}(\boldsymbol{\omega}^{t-1}) \right\} \quad \forall t \in \tau_{-0}. \tag{2.7}$$

Notice that Y^t can be considered as the graph of the multifunction X^t over Ω^t or as the graph of X_t over Z^t, as sketched in Fig. 2.3. Moreover, as indicated in the illustration, Y^t lies within the *level set* $\text{lev}_{\leq 0}\, \boldsymbol{f}_t$, where

$$\text{lev}_{\leq 0}\, \boldsymbol{f}_t := \{ (\boldsymbol{x}^t, \boldsymbol{\omega}^t) \,|\, \boldsymbol{f}_t(\boldsymbol{x}^t, \boldsymbol{\omega}^t) \leq \boldsymbol{0} \}.$$

The generalized feasible sets Y^t and Z^t are completely determined by the constraint functions, and their importance will become clear in the next sections.

2.4 Static and Dynamic Version of a Stochastic Program

In stochastic programming it is usually assumed that decisions are selected in order to maximize the expected value of some time-separable objective function. The intertemporal contributions to the objective, $\{\rho_t\}_{t\in\tau}$, depend on the outcome and decision history and can be interpreted as a sequence of *profit functions*. As a minimal requirement, the mappings $\rho_t : \mathbb{R}^{n^t} \times \mathbb{R}^{M^t} \to \mathbb{R}$ $(t \in \tau)$ must be Borel measurable. Collecting the above definitions, the *static version* of a general *(non-linear) multistage stochastic program* can be formulated as follows:

$$\sup_{\tilde{\boldsymbol{x}}\in\mathcal{N}_n} \int_{\Omega} \left[\sum_{t=0}^{T} \rho_t(\tilde{\boldsymbol{x}}^t(\boldsymbol{\omega}^t), \boldsymbol{\omega}^t) \right] dP(\boldsymbol{\omega}) \tag{2.8}$$
$$\text{s.t.} \quad \boldsymbol{f}_t(\tilde{\boldsymbol{x}}^t(\tilde{\boldsymbol{\omega}}^t), \tilde{\boldsymbol{\omega}}^t) \le \mathbf{0} \quad P\text{-a.s.} \quad t \in \tau.$$

For theoretical considerations it is sometimes useful to bring the explicit restrictions to the objective function. To this end we introduce the *effective profit functions* $\{\hat{\rho}_t\}_{t\in\tau}$, which are defined as extended-real-valued mappings:

$$\hat{\rho}_t(\boldsymbol{x}^t, \boldsymbol{\omega}^t) := \begin{cases} \rho_t(\boldsymbol{x}^t, \boldsymbol{\omega}^t) & \text{for } \boldsymbol{f}_t(\boldsymbol{x}^t, \boldsymbol{\omega}^t) \le \mathbf{0}, \\ -\infty & \text{else.} \end{cases} \tag{2.9}$$

With (2.9) the stochastic program (2.8) reduces to

$$\sup_{\tilde{\boldsymbol{x}}\in\mathcal{N}_n} \int_{\Omega} \left[\sum_{t=0}^{T} \hat{\rho}_t(\tilde{\boldsymbol{x}}^t(\boldsymbol{\omega}^t), \boldsymbol{\omega}^t) \right] dP(\boldsymbol{\omega}). \tag{2.10}$$

Notice that (2.10) and (2.8) are indeed equivalent since the explicit constraints in (2.8) are assumed to hold almost surely. On the other hand, allowing for policy functions which violate the restrictions on P-null sets has no undesirable effect on the optimal solution. In fact, after a suitable redefinition on a set of measure zero, any feasible policy function strictly satisfies the explicit restrictions, and any such redefinition leaves the objective function value unchanged.

Sometimes it is more comfortable to work with the *dynamic version* of a given multistage stochastic program, which is defined recursively. The parametric family of stage T subproblems is given through

$$\Phi_T(\boldsymbol{x}^{T-1}, \boldsymbol{\omega}^T) := \sup_{\boldsymbol{x}_T\in\mathbb{R}^{n_T}} \hat{\rho}_T(\boldsymbol{x}^T, \boldsymbol{\omega}^T), \tag{2.11a}$$

whereas the subordinate maximization problems for $t = T-1, \dots, 1$ depend on the optimal value functions of the subsequent stages, respectively.

$$\Phi_t(\boldsymbol{x}^{t-1}, \boldsymbol{\omega}^t) := \sup_{\boldsymbol{x}_t\in\mathbb{R}^{n_t}} \hat{\rho}_t(\boldsymbol{x}^t, \boldsymbol{\omega}^t) + \int_{\Omega_{t+1}} \Phi_{t+1}(\boldsymbol{x}^t, \boldsymbol{\omega}^{t+1}) dP_{t+1}(\omega_{t+1}|\boldsymbol{\omega}^t) \tag{2.11b}$$

The optimal value function of the subordinate zero stage problem only depends on the random variable $\tilde{\omega}_0$, which is usually assumed deterministic.

$$\Phi_0(\omega_0) := \sup_{x_0 \in \mathbb{R}^{n_0}} \hat{\rho}_0(x_0, \omega_0) + \int_{\Omega_1} \Phi_1(x_0, \omega_0) dP_1(\omega_1 | \omega_0) \tag{2.11c}$$

Most importantly, the optimal value $(E\Phi_0) := \int \Phi_0 dP_0$ of the dynamic program (2.11) reduces to the unconditional expectation of the value function Φ_0. If $\tilde{\omega}_0$ is deterministic, the calculation of $(E\Phi_0)$ reduces to the evaluation of Φ_0 at one single point.

In the above problem formulation, P_t denotes the regular conditional probability distribution of $\tilde{\omega}_t$ given $\tilde{\omega}^{t-1}$ (for $t \in \tau_{-0}$), and P_0 stands for the marginal distribution of $\tilde{\omega}_0$. Notice that the probability measure $P_t(\cdot|\omega^{t-1})$ does not depend on the decision history x^{t-1}, implying that the uncertain impacts are completely exogenous and can not be influenced by the decision maker. It should be remarked that the maximization in problem (2.8) is performed over all feasible policy functions $\tilde{x} \in \mathcal{N}_n$, whereas the recursive formulation (2.11) represents a large collection of simpler optimization problems over finite-dimensional Euclidean spaces. The optimal value function of stage t characterizes an extended-real-valued mapping $\Phi_t : \mathbb{R}^{n^{t-1}} \times \mathbb{R}^{M^t} \to [-\infty, \infty]$, which is usually referred to as *recourse function* in literature [10,61]. Furthermore, in order to simplify notation, we define the *expectation functional*

$$(E_t\Phi_{t+1})(x^t, \omega^t) := \int_{\Omega_{t+1}} \Phi_{t+1}(x^t, \omega^{t+1}) dP_{t+1}(\omega_{t+1}|\omega^t) \tag{2.12}$$

as the expectation of the recourse function Φ_{t+1} conditional on the information available at stage t, $t \in \tau_{-T}$.

Before choosing an action at stage t, the decision maker knows the outcome and decision history $(x^{t-1}, \omega^t) \in Z^t$. Vectors in the complement of Z^t correspond to sequences of impossible outcomes or forbidden actions. In principle, they need not be considered. Thus, only the restriction of Φ_t to Z^t has physical meaning, and Z^t shall be referred to as the *natural domain* of the recourse function Φ_t. Moreover, it is important to realize that changes of ρ_t on the complement of Y^t do not influence the restriction of Φ_s to Z^s, $s = 0, \ldots, t$. Therefore, Y^t can conveniently be interpreted as the *natural domain* of the profit function ρ_t and the expectation functional $(E_t\Phi_{t+1})$.

It is intuitively appealing that the static and dynamic versions of a multistage stochastic program are equivalent in the sense that their optimal values coincide. However, without adequate regularity conditions it is not even clear whether the integrals in (2.11b) and (2.11c) are well-defined. Therefore, in a first step, we must care about the existence of (2.8) and (2.11). Once well-definedness is established, the static and dynamic versions can be shown to be equivalent and solvable.

For the precise formulation of appropriate regularity conditions we need the following conventions.

Definition 2.1. *We call* $\{\tilde{\omega}_t\}_{t\in\mathcal{T}}$ *a nonlinear autoregressive process if for every index* $t \in \mathcal{T}_{-0}$ *there is a random vector* $\tilde{\varepsilon}_t$ *with induced probability space* $(\mathcal{E}_t, \mathcal{B}(\mathcal{E}_t), Q_t)$ *and a transformation function* $H_t : \Omega^{t-1} \times \mathcal{E}_t \to \Omega_t$ *such that*

$$\tilde{\omega}_t = H_t(\tilde{\omega}^{t-1}, \tilde{\varepsilon}_t).$$

The initial data $\tilde{\omega}_0$ *and the disturbances* $\{\tilde{\varepsilon}_t\}_{t\in\mathcal{T}_{-0}}$ *are mutually independent, and for each* $t \in \mathcal{T}_{-0}$ *we postulate:*

- $\mathcal{E}_t$ *is a non-empty Borel subset of a finite-dimensional Euclidean space;*
- H_t *is a Borel measurable map;*
- Q_t *is a regular probability measure on* $(\mathcal{E}_t, \mathcal{B}(\mathcal{E}_t))$.

Definition 2.1 is inspired by a similar concept in the field of nonlinear time series analysis [105, Chap. 3]. Note that if $\{\tilde{\omega}_t\}_{t\in\mathcal{T}}$ follows a nonlinear autoregressive process, then the conditional probability distribution P_t is defined through

$$P_t(A|\omega^{t-1}) = Q_t(\{\varepsilon_t | H_t(\omega^{t-1}, \varepsilon_t) \in A\}) \quad \forall\, A \in \mathcal{B}(\Omega_t),\ \omega^{t-1} \in \Omega^{t-1}.$$

Definition 2.2. *A set-valued mapping* $X : \Omega \subset \mathbb{R}^M \to 2^{\mathbb{R}^n}$ *is said to be*

(i) upper semicontinuous (usc) at $\bar{\omega}$ *if for each open set* U *with* $X(\bar{\omega}) \subset U$ *the set* $\{\omega | X(\omega) \subset U\}$ *is a neighborhood of* $\bar{\omega}$*;*

(ii) lower semicontinuous (lsc) at $\bar{\omega}$ *if for each open set* U *with* $X(\bar{\omega}) \cap U \neq \emptyset$ *the set* $\{\omega | X(\omega) \cap U \neq \emptyset\}$ *is a neighborhood of* $\bar{\omega}$*;*

(iii) continuous at $\bar{\omega}$ *if it is both usc and lsc at* $\bar{\omega}$*;*

(iv) usc (lsc, continuous) on Ω *if it is usc (lsc, continuous) at* $\bar{\omega}$ *for every* $\bar{\omega} \in \Omega$*;*

(v) bounded on Ω *if the image* $X(\Omega)$ *is a bounded subset of* $\mathbb{R}^n$.

Definition 2.2 is consistent with Border [13, definition 11.3]. In that reference, however, semicontinuity of multifunctions is referred to as 'hemicontinuity', in order to avoid confusion with semicontinuity of ordinary functions. For more details about continuity properties of multifunctions, we refer to the concise treatment by Rockafellar and Wets [97, Chap. 5].

Returning to the stochastic program (2.8), we will impose the following regularity conditions:

(A1) Ω_t is compact and covers the support of $\tilde{\omega}_t$ for all $t \in \mathcal{T}$;
(A2) the profit function ρ_t is usc and bounded on Y^t for all $t \in \mathcal{T}$;
(A3) the feasible set mapping X_t is usc and bounded on Z^t for all $t \in \mathcal{T}$;
(A4) the random data $\{\tilde{\omega}_t\}_{t\in\mathcal{T}}$ follows a nonlinear autoregressive process; H_t is a *Carathéodory map*, i.e. it is continuous in ω^{t-1} and (Borel) measurable in ε_t, while $\mathcal{E}_t$ is compact for all $t \in \mathcal{T}_{-0}$;
(A5) the multifunction X_t is non-empty-closed-valued on Z^t for all $t \in \mathcal{T}$.

The conditions (A1), (A2), (A3), and (A4) are unproblematic for many practical applications (see also proposition 2.3 below). Assumption (A5) ensures non-anticipativity of the constraint multifunction. The following proposition suggests a simple characterization of closed-valued feasible set mappings which are usc and bounded.

Proposition 2.3. *Given the condition (A1), assumption (A3) is equivalent to compactness of the generalized feasible sets* $\{Y^t\}_{t\in\mathcal{T}}$ *and* $\{Z^t\}_{t\in\mathcal{T}}$.

Proof: First, we prove necessity by induction on t. The basis step is trivial since $Z^0 = \Omega_0$ is compact as implied by assumption (A1). Next, assume that Z^t is compact for some $t \in \mathcal{T}$. Then, assumption (A3) postulates boundedness of $X_t(Z^t)$. Thus, the generalized feasible set

$$Y^t = \operatorname{graph} X_t|_{Z^t} \subset Z^t \times X_t(Z^t)$$

is bounded. Moreover, as X_t is usc and closed-valued, we may invoke proposition 11.9 (a) of Border [13] to prove closedness of Y^t. Boundedness and closedness entail compactness of Y^t. If $t < T$, by assumption (A1) we may then conclude that $Z^{t+1} = Y^t \times \Omega_{t+1}$ is compact, too. This observation completes the induction step.

Next, we prove sufficiency. By assumption, the graph of the multifunction X_t over Z^t is compact. Therefore, the image of X_t over Z^t is bounded, and X_t is usc due to [13, proposition 11.9 (b)]. Hence, condition (A3) follows. □

Corollary 2.4. *Given condition (A1), assumption (A3) holds true if the constraint function* $\boldsymbol{f}_t$ *is continuous, and the set* $\operatorname{lev}_{\leq 0} \boldsymbol{f}_t$ *is bounded for all* $t \in \mathcal{T}$.

Proposition 2.5. *Under the assumptions (A1)–(A5) the recourse function* Φ_t *is usc and bounded on* Z^t *for all* $t \in \mathcal{T}$. *A fortiori, the expectation functional* $(E_t\Phi_{t+1})$ *is well-defined, finite, and usc on* Y^t *for* $t \in \mathcal{T}_{-T}$, *and* $(E\Phi_0)$ *is finite.*

Proof. Denote by $\underline{\kappa}_t$ and $\overline{\kappa}_t$ finite lower and upper bounds of ρ_t on Y^t. Such bounds exist by assumption (A2). Then, the claim is proved by backward induction with respect to t. By condition (A2), the effective profit function $\hat{\rho}_T$ is usc and bounded on Y^T. For $(\boldsymbol{x}^{T-1}, \boldsymbol{\omega}^T) \in Z^T$ the recourse function of stage T is given by

$$\Phi_T(\boldsymbol{x}^{T-1}, \boldsymbol{\omega}^T) = \sup_{\boldsymbol{x}_T \in \mathbb{R}^{n_T}} \hat{\rho}_T(\boldsymbol{x}^T, \boldsymbol{\omega}^T) = \sup_{\boldsymbol{x}_T \in X_T(Z^T)} \hat{\rho}_T(\boldsymbol{x}^T, \boldsymbol{\omega}^T).$$

Thus, by [97, theorem 1.17] the optimal value function Φ_T is usc on Z^T since $X_T(Z^T)$ is compact. Notice that $X_T(Z^T)$ is the projection of Y^T on $\mathbb{R}^{n_T}$, which is compact by proposition 2.3. Assumption (A5) and the definition of the recourse function Φ_T directly yield the inequality $\underline{\kappa}_T \leq \Phi_T \leq \overline{\kappa}_T$ on Z^T. Thus, the basis step is established.

Assume next that the recourse function Φ_{t+1} is usc and bounded on Z^{t+1}, i.e. assume $\sum_{s=t+1}^{T} \underline{\kappa}_s \leq \Phi_{t+1} \leq \sum_{s=t+1}^{T} \overline{\kappa}_s$. Upper semicontinuity entails Borel measurability of the mapping $\boldsymbol{\omega}_{t+1} \mapsto \Phi_{t+1}(\boldsymbol{x}^t, \boldsymbol{\omega}^{t+1})$. Consequently, the expectation functional $(E_t\Phi_{t+1})$ is well-defined. Choosing an arbitrary convergent sequence $\{(\boldsymbol{x}_k^t, \boldsymbol{\omega}_k^t)\}_{k \in \mathbb{N}}$ in Y^t with a limiting point $(\boldsymbol{x}^t, \boldsymbol{\omega}^t) \in Y^t$ we find

$$\begin{aligned}
&\limsup_{k\to\infty} (E_t\Phi_{t+1})(\boldsymbol{x}_k^t, \boldsymbol{\omega}_k^t) \\
&= \limsup_{k\to\infty} \int_{\Omega_{t+1}} \Phi_{t+1}(\boldsymbol{x}_k^t, \boldsymbol{\omega}_{t+1}, \boldsymbol{\omega}_k^t) dP_{t+1}(\boldsymbol{\omega}_{t+1}|\boldsymbol{\omega}_k^t) \\
&= \limsup_{k\to\infty} \int_{\mathcal{E}_{t+1}} \Phi_{t+1}(\boldsymbol{x}_k^t, H_{t+1}(\boldsymbol{\omega}_k^t, \boldsymbol{\varepsilon}_{t+1}), \boldsymbol{\omega}_k^t) dQ_{t+1}(\boldsymbol{\varepsilon}_{t+1}) \\
&\leq \int_{\mathcal{E}_{t+1}} \limsup_{k\to\infty} \Phi_{t+1}(\boldsymbol{x}_k^t, H_{t+1}(\boldsymbol{\omega}_k^t, \boldsymbol{\varepsilon}_{t+1}), \boldsymbol{\omega}_k^t) dQ_{t+1}(\boldsymbol{\varepsilon}_{t+1}) \\
&\leq \int_{\mathcal{E}_{t+1}} \Phi_{t+1}(\boldsymbol{x}^t, H_{t+1}(\boldsymbol{\omega}^t, \boldsymbol{\varepsilon}_{t+1}), \boldsymbol{\omega}^t) dQ_{t+1}(\boldsymbol{\varepsilon}_{t+1}) \\
&= (E_t\Phi_{t+1})(\boldsymbol{x}^t, \boldsymbol{\omega}^t).
\end{aligned}$$

The inequalities hold because of Fatou's lemma (which applies since Φ_{t+1} is bounded from above) and the upper semicontinuity of Φ_{t+1} in conjunction with the continuity of $\boldsymbol{\omega}^t \mapsto H_{t+1}(\boldsymbol{\omega}^t, \boldsymbol{\varepsilon}_{t+1})$. Therefore, $(E_t\Phi_{t+1})$ is usc on Y^t. Moreover, the definition of the expectation functional and the induction hypothesis imply that $\sum_{s=t+1}^{T} \underline{\kappa}_s \leq (E_t\Phi_{t+1}) \leq \sum_{s=t+1}^{T} \overline{\kappa}_s$ on Y^t.

By assumption, the effective profit function $\hat{\rho}_t$ is usc and bounded on Y^t. For $(\boldsymbol{x}^{t-1}, \boldsymbol{\omega}^t) \in Z^t$ the recourse function of stage t is given by

$$\begin{aligned}
\Phi_t(\boldsymbol{x}^{t-1}, \boldsymbol{\omega}^t) &= \sup_{\boldsymbol{x}_t \in \mathbb{R}^{n_t}} \hat{\rho}_t(\boldsymbol{x}^t, \boldsymbol{\omega}^t) + (E_t\Phi_{t+1})(\boldsymbol{x}^t, \boldsymbol{\omega}^t) \\
&= \sup_{\boldsymbol{x}_t \in X_t(Z^t)} \hat{\rho}_t(\boldsymbol{x}^t, \boldsymbol{\omega}^t) + (E_t\Phi_{t+1})(\boldsymbol{x}^t, \boldsymbol{\omega}^t).
\end{aligned}$$

Thus, by [97, theorem 1.17] Φ_t is usc on Z^t since $X_t(Z^t)$ is compact (notice that $X_t(Z^t)$ is the projection of Y^t on $\mathbb{R}^{n_t}$, which is compact by proposition 2.3; see also Fig. 2.3). Assumption (A5) and the definition of the recourse function Φ_t directly yield the inequality

$$\sum_{s=t}^{T} \underline{\kappa}_s \leq \Phi_t(\boldsymbol{x}^{t-1}, \boldsymbol{\omega}^t) \leq \sum_{s=t}^{T} \overline{\kappa}_s \quad \text{on } Z^t.$$

By induction, Φ_t is usc and bounded for every $t \in \tau$, and $(E_t\Phi_{t+1})$ is usc and bounded for $t \in \tau_{-T}$. Finiteness of the optimal value $(E\Phi_0)$ follows from an elementary argument. □

Theorem 2.6. *Under the assumptions (A1)–(A5) the static and dynamic versions of a multistage stochastic program are both solvable, and the optimal values coincide.*

Proof. Let $(\Omega^t, \mathcal{B}(\Omega^t), P^t)$ be the marginal probability space associated with the random variables appearing in the first t stages. Moreover, define an auxiliary function

$$p_t(\boldsymbol{x}^t, \boldsymbol{\omega}^t) := \begin{cases} \sum_{s=0}^{t} \hat{\rho}_s(\boldsymbol{x}^s, \boldsymbol{\omega}^s) + \big(E_t\Phi_{t+1}\big)(\boldsymbol{x}^t, \boldsymbol{\omega}^t) & \text{for } t < T, \\ \sum_{s=0}^{T} \hat{\rho}_s(\boldsymbol{x}^s, \boldsymbol{\omega}^s) & \text{for } t = T. \end{cases}$$

We use induction by t to prove that

$$\sup_{\tilde{\boldsymbol{x}}^t \in \mathcal{N}_{n^t}} \int_{\Omega^t} p_t(\tilde{\boldsymbol{x}}^t(\boldsymbol{\omega}^t), \boldsymbol{\omega}^t)\, dP^t(\boldsymbol{\omega}^t) \tag{2.13}$$

is solvable on the space of non-anticipative policies[5] up to time t, $\mathcal{N}_{n^t} := \times_{s=0}^{t} \mathcal{N}_{n_s}$, and that the optimal value is given by $(E\Phi_0)$. Beginning at stage $t = 0$, we may reformulate (2.11c) and calculate the unconditional expectation value to obtain

$$(E\Phi_0) = \int_{\Omega_0} \sup_{\boldsymbol{x}_0 \in \mathbb{R}^{n_0}} p_0(\boldsymbol{x}_0, \boldsymbol{\omega}_0)\, dP_0(\boldsymbol{\omega}_0).$$

By proposition 2.5 the auxiliary map p_0 is usc and finite on Y^0. Furthermore, by the definition of the effective profit functions we have $p_0(\boldsymbol{x}_0, \boldsymbol{\omega}_0) = -\infty$ for $(\boldsymbol{x}_0, \boldsymbol{\omega}_0) \notin Y_0$. Thus, p_0 is usc on $\mathbb{R}^{n_0} \times \Omega_0$ and its effective domain is given by $\operatorname{dom} p_0 = Y^0$. These results can be used to show that p_0 is a *normal integrand* in the sense of [91]. Concretely speaking, we must prove that the hypograph multifunction $\boldsymbol{\omega}_0 \mapsto \mathcal{H}_{p_0}(\boldsymbol{\omega}_0) := \operatorname{hypo} p_0(\cdot, \boldsymbol{\omega}_0)$ is measurable and closed-valued. Obviously, $\mathcal{H}_{p_0}$ is closed-valued as a direct consequence of the upper

[5] Note that $\mathcal{N}_{n_t}$ is naturally isomorphic to $\mathcal{L}^\infty(\Omega^t, \mathcal{B}(\Omega^t), P^t; \mathbb{R}^{n_t})$. Hence, these spaces will be identified in the remainder.

semicontinuity of the mapping $x_0 \mapsto p_0(x_0, \omega_0)$. To prove measurability, it is sufficient to verify that

$$\mathcal{H}_{p_0}^{-1}(C) := \{\omega_0 \in \Omega_0 \,|\, \mathcal{H}_{p_0}(\omega_0) \cap C \neq \emptyset\}$$

is Borel measurable for all compact sets $C \subset \mathbb{R}^{n_0}$ [91, proposition 1A]. Recall that the hypograph of p_0, which coincides with the graph of the multifunction $\mathcal{H}_{p_0}$, is closed since p_0 is usc. In addition, notice that $\mathcal{H}_{p_0}^{-1}(C)$ is given by the projection of $(C \times \Omega_0) \cap \operatorname{graph} \mathcal{H}_{p_0}$ on Ω_0. Then, compactness of $C \times \Omega_0$ implies that $(C \times \Omega_0) \cap \operatorname{graph} \mathcal{H}_{p_0}$ is compact, and $\mathcal{H}_{p_0}^{-1}(C)$ is compact and Borel measurable as a projection of a compact set. Therefore, p_0 is a normal integrand on $\mathbb{R}^{n_0} \times \Omega_0$.

For all $\tilde{x}_0 \in \mathcal{N}_{n_0}$ we have

$$p_0(\tilde{x}_0(\omega_0), \omega_0) \leq \sup_{x_0 \in \mathbb{R}^{n_0}} p_0(x_0, \omega_0). \tag{2.14}$$

The supremum on the right-hand side of (2.14) is attained for all $\omega_0 \in \Omega_0$ since p_0 is usc and its effective domain is compact. Consequently, the multifunction

$$\Gamma(\omega_0) := \arg\max \left\{ p_0(x_0, \omega_0) \,\middle|\, x_0 \in \mathbb{R}^{n_0} \right\}$$

is non-empty-closed-valued and measurable on the entire space Ω_0. Measurability follows from [91, theorem 2K]. Moreover, by [91, corollary 1C] there exists a measurable selector $\tilde{x}_{\mathrm{opt},0} : \Omega_0 \to \mathbb{R}^{n_0}$ such that

$$\tilde{x}_{\mathrm{opt},0}(\omega_0) \in \Gamma(\omega_0) \quad \forall \omega_0 \in \Omega_0.$$

Since $\Gamma(\omega_0) \subset X_0(Z^0)$ and $X_0(Z^0)$ is compact, we may conclude that $\tilde{x}_{\mathrm{opt},0}$ is essentially bounded. The estimate (2.14) together with the existence of an optimal policy $\tilde{x}_{\mathrm{opt},0} \in \mathcal{N}_{n_0}$ implies

$$(E\Phi_0) = \sup_{\tilde{x}_0 \in \mathcal{N}_{n_0}} \int_{\Omega_0} p_0(\tilde{x}_0(\omega_0), \omega_0) \, dP_0(\omega_0). \tag{2.15}$$

This proves that the optimal value of (2.13) is given by $(E\Phi_0)$, and thus the basis step is established.

Next, assume that the claim holds for $t-1$ and that $\tilde{x}_{\mathrm{opt}}^{t-1} \in \mathcal{N}_{n^{t-1}}$ is an optimal policy. Substituting the definitions of the expectation functional $(E_{t-1}\Phi_t)$ and the recourse function Φ_t into the induction hypothesis, applying the *generalized Fubini theorem*,[6] and rearranging terms we obtain

[6]For every $\mathcal{F}^t$-measurable function $\varphi : \Omega^t \to [-\infty, \infty]$ we have the identity (for a proof see e.g. [2, theorem 2.6.4])

$$\int_{\Omega^t} \varphi(\omega^t) \, dP^t(\omega^t) = \int_{\Omega^{t-1}} \int_{\Omega_t} \varphi(\omega^t) \, dP_t(\omega_t | \omega^{t-1}) dP^{t-1}(\omega^{t-1}).$$

$$(E\Phi_0) = \sup_{\tilde{\boldsymbol{x}}^{t-1}\in\mathcal{N}_{n^{t-1}}} \int_{\Omega^{t-1}} p_{t-1}(\tilde{\boldsymbol{x}}^{t-1}(\boldsymbol{\omega}^{t-1}), \boldsymbol{\omega}^{t-1})\, dP^{t-1}(\boldsymbol{\omega}^{t-1})$$
$$= \sup_{\tilde{\boldsymbol{x}}^{t-1}\in\mathcal{N}_{n^{t-1}}} \int_{\Omega^t} \sup_{\boldsymbol{x}_t\in\mathbb{R}^{n_t}} p_t(\boldsymbol{x}_t, \tilde{\boldsymbol{x}}^{t-1}(\boldsymbol{\omega}^{t-1}), \boldsymbol{\omega}^t)\, dP^t(\boldsymbol{\omega}^t).$$

Recall that p_t is usc and finite on its effective domain Y^t, which is compact. An argument parallel to the one in the basis step proves that p_t is a normal integrand on $\mathbb{R}^{n^t} \times \Omega^t$ in the sense of [91]. For the further argumentation we fix an arbitrary policy $\tilde{\boldsymbol{x}}^{t-1} \in \mathcal{N}_{n^{t-1}}$. Then, the mapping

$$(\boldsymbol{x}_t, \boldsymbol{\omega}^t) \mapsto p_t(\tilde{\boldsymbol{x}}^{t-1}(\boldsymbol{\omega}^{t-1}), \boldsymbol{x}_t, \boldsymbol{\omega}^t)$$

is a normal integrand on $\mathbb{R}^{n_t} \times \Omega^t$ due to [91, corollary 2P]. For all $\tilde{\boldsymbol{x}}_t \in \mathcal{N}_{n_t}$ we have

$$p_t(\tilde{\boldsymbol{x}}^{t-1}(\boldsymbol{\omega}^{t-1}), \tilde{\boldsymbol{x}}_t(\boldsymbol{\omega}^t), \boldsymbol{\omega}^t) \leq \sup_{\boldsymbol{x}_t\in\mathbb{R}^{n_t}} p_t(\tilde{\boldsymbol{x}}^{t-1}(\boldsymbol{\omega}^{t-1}), \boldsymbol{x}_t, \boldsymbol{\omega}^t). \tag{2.16}$$

The supremum on the right-hand side of (2.16) is attained for all $\boldsymbol{\omega}^t \in \Omega^t$ since p_t is usc and its effective domain is compact. Consequently, the multifunction

$$\Gamma(\boldsymbol{\omega}^t) := \arg\max \left\{ p_t(\tilde{\boldsymbol{x}}^{t-1}(\boldsymbol{\omega}^{t-1}), \boldsymbol{x}_t, \boldsymbol{\omega}^t) \,\middle|\, \boldsymbol{x}_t \in \mathbb{R}^{n_t} \right\}$$

is non-empty-closed-valued and measurable on the entire space Ω^t. Measurability follows from [91, theorem 2K]. Moreover, by [91, corollary 1C] there exists a measurable selector[7] $\tilde{\boldsymbol{x}}_t^\star : \Omega^t \to \mathbb{R}^{n_t}$ such that

$$\tilde{\boldsymbol{x}}_t^\star(\boldsymbol{\omega}^t) \in \Gamma(\boldsymbol{\omega}^t) \quad \forall \boldsymbol{\omega}^t \in \Omega^t.$$

Since $\Gamma(\boldsymbol{\omega}^t) \subset X_t(Z^t)$ and $X_t(Z^t)$ is compact, we may conclude that $\tilde{\boldsymbol{x}}_t^\star$ is essentially bounded. The estimate (2.16) together with the existence of an optimal policy $\tilde{\boldsymbol{x}}_t^\star \in \mathcal{N}_{n_t}$ implies

$$(E\Phi_0) = \sup_{\tilde{\boldsymbol{x}}^{t-1}\in\mathcal{N}_{n^{t-1}}} \int_{\Omega^t} \sup_{\boldsymbol{x}_t\in\mathbb{R}^{n_t}} p_t(\tilde{\boldsymbol{x}}^{t-1}(\boldsymbol{\omega}^{t-1}), \boldsymbol{x}_t, \boldsymbol{\omega}^t)\, dP^t(\boldsymbol{\omega}^t)$$
$$= \sup_{\tilde{\boldsymbol{x}}^{t-1}\in\mathcal{N}_{n^{t-1}}} \sup_{\tilde{\boldsymbol{x}}_t\in\mathcal{N}_{n_t}} \int_{\Omega^t} p_t(\tilde{\boldsymbol{x}}^{t-1}(\boldsymbol{\omega}^{t-1}), \tilde{\boldsymbol{x}}_t(\boldsymbol{\omega}^t), \boldsymbol{\omega}^t)\, dP^t(\boldsymbol{\omega}^t)$$
$$= \sup_{\tilde{\boldsymbol{x}}^t\in\mathcal{N}_{n^t}} \int_{\Omega^t} p_t(\tilde{\boldsymbol{x}}^t(\boldsymbol{\omega}^t), \boldsymbol{\omega}^t)\, dP^t(\boldsymbol{\omega}^t). \tag{2.17}$$

The last step proves that the optimal value of (2.13) is given by $(E\Phi_0)$. In the third line of (2.17) the two 'sup'-operators may be combined because of proposition A.4 in the appendix.

[7]Notice that $\tilde{\boldsymbol{x}}_t^\star$ depends on the choice of $\tilde{\boldsymbol{x}}^{t-1} \in \mathcal{N}_{n^{t-1}}$.

In order to establish solvability, consider the policy $\tilde{\boldsymbol{x}}_{\text{opt}}^{t-1} \in \mathcal{N}_{n^{t-1}}$ given by the induction hypothesis. Repeating the arguments of the previous paragraph, we can show that there is a measurable function $\tilde{\boldsymbol{x}}_{\text{opt},t} \in \mathcal{N}_{n_t}$ such that

$$\tilde{\boldsymbol{x}}_{\text{opt},t}(\boldsymbol{\omega}^t) \in \arg\max \left\{ p_t(\tilde{\boldsymbol{x}}_{\text{opt}}^{t-1}(\boldsymbol{\omega}^{t-1}), \boldsymbol{x}_t, \boldsymbol{\omega}^t) \,\middle|\, \boldsymbol{x}_t \in \mathbb{R}^{n_t} \right\} \quad \forall \boldsymbol{\omega}^t \in \Omega^t.$$

Thus, it can easily be verified that $\tilde{\boldsymbol{x}}_{\text{opt}}^t := (\tilde{\boldsymbol{x}}_{\text{opt}}^{t-1}, \tilde{\boldsymbol{x}}_{\text{opt},t})$ is a maximizer of (2.13) on the space of non-anticipative policies up to time t, i.e.

$$(E\Phi_0) = \int_{\Omega^t} p_t(\tilde{\boldsymbol{x}}_{\text{opt}}^t(\boldsymbol{\omega}^t), \boldsymbol{\omega}^t)\, dP^t(\boldsymbol{\omega}^t).$$

For $t = T$ (2.13) reduces to the static version of the stochastic program (2.8) and its optimal value coincides with $(E\Phi_0)$. In particular, (2.8) is solvable, and thus the claim is proved. □

Proposition 2.5 and theorem 2.6 are inspired by [92] and the dynamic programming framework developed in Klein Haneveld [64, Sect. 6.3].

Many decision problems of practical relevance can be formulated as multistage stochastic programs satisfying the regularity conditions (A1)–(A5). However, sometimes it proves useful to work with a set of slightly more restrictive regularity conditions, which impose stronger structural properties on the recourse functions.

(A1)' = (A1);
(A2)' the profit function ρ_t is continuous on Y^t for all $t \in \tau$;
(A3)' the feasible set mapping X_t is continuous and bounded on Z^t for all $t \in \tau$;
(A4)' the random data $\{\tilde{\omega}_t\}_{t\in\tau}$ follows a nonlinear autoregressive process; H_t is a continuous function, and $\mathcal{E}_t$ is compact for all $t \in \tau_{-0}$;
(A5)' = (A5).

These regularity conditions are obtained from the less restrictive conditions (A1)–(A5) if we replace upper semicontinuity by ordinary continuity in (A2) and (A3) and if we require the transformation function H_t to be continuous in (A4). One can easily verify that the conditions (A1)–(A5) are implied by (A1)'–(A5)'. Notice that boundedness of the profit functions need not be postulated explicitly in (A2)'. Instead, it follows directly from the Weierstrass maximum principle. It is also clear that the statements of the propositions 2.5 and 2.6 as well as necessity in proposition 2.3 still hold true under the regularity conditions (A1)'–(A5)'. In addition, one can now prove continuity of the recourse functions.

Proposition 2.7. *Under the assumptions (A1)'–(A5)' the recourse function Φ_t is continuous on Z^t for all $t \in \tau$. Moreover, the expectation functional $(E_t\Phi_{t+1})$ is continuous on Y^t for all $t \in \tau_{-T}$.*

Proof. The claim is proved by backward induction with respect to t. For any outcome and decision history $(\boldsymbol{x}^{T-1}, \boldsymbol{\omega}^T) \in Z^T$ the stage T recourse function can be written as

$$\Phi_T(\boldsymbol{x}^{T-1}, \boldsymbol{\omega}^T) = \sup \left\{ \rho_T(\boldsymbol{x}^T, \boldsymbol{\omega}^T) \,\middle|\, \boldsymbol{x}_T \in X_T(\boldsymbol{x}^{T-1}, \boldsymbol{\omega}^T) \right\}.$$

By assumption (A2)' the objective function of this maximization problem is continuous on Y^T. Moreover, assumption (A3)' requires the feasible set mapping X_T to be continuous on Z^T. Thus, by Berge's theorem [13, theorem 12.1] Φ_T is continuous on Z^T. Hence, the basis step is proved.

Assume now that the recourse function Φ_{t+1} is continuous on Z^{t+1}. Then, choosing an arbitrary convergent sequence $\{(\boldsymbol{x}_k^t, \boldsymbol{\omega}_k^t)\}_{k \in \mathbb{N}}$ in Y^t with a limiting point $(\boldsymbol{x}^t, \boldsymbol{\omega}^t) \in Y^t$ we find

$$\begin{aligned}
&\lim_{k\to\infty} (E_t\Phi_{t+1})(\boldsymbol{x}_k^t, \boldsymbol{\omega}_k^t) \\
&= \lim_{k\to\infty} \int_{\mathcal{E}_{t+1}} \Phi_{t+1}(\boldsymbol{x}_k^t, H_{t+1}(\boldsymbol{\omega}_k^t, \varepsilon_{t+1}), \boldsymbol{\omega}_k^t) dQ_{t+1}(\varepsilon_{t+1}) \\
&= \int_{\mathcal{E}_{t+1}} \lim_{k\to\infty} \Phi_{t+1}(\boldsymbol{x}_k^t, H_{t+1}(\boldsymbol{\omega}_k^t, \varepsilon_{t+1}), \boldsymbol{\omega}_k^t) dQ_{t+1}(\varepsilon_{t+1}) \\
&= \int_{\mathcal{E}_{t+1}} \Phi_{t+1}(\boldsymbol{x}^t, H_{t+1}(\boldsymbol{\omega}^t, \varepsilon_{t+1}), \boldsymbol{\omega}^t) dQ_{t+1}(\varepsilon_{t+1}) \\
&= (E_t\Phi_{t+1})(\boldsymbol{x}^t, \boldsymbol{\omega}^t).
\end{aligned}$$

The second and the third equalities hold because of the dominated convergence theorem (which applies since Φ_{t+1} is bounded and the integration region is compact) and continuity of the integrand. Therefore, $(E_t\Phi_{t+1})$ is continuous on its natural domain Y^t.

For $(\boldsymbol{x}^{t-1}, \boldsymbol{\omega}^t) \in Z^t$ the stage t recourse function is representable as

$$\begin{aligned}
\Phi_t(\boldsymbol{x}^{t-1}, \boldsymbol{\omega}^t) = \sup_{\boldsymbol{x}_t \in \mathbb{R}^{n_t}} \ & \rho_t(\boldsymbol{x}^t, \boldsymbol{\omega}^t) + (E_t\Phi_{t+1})(\boldsymbol{x}^t, \boldsymbol{\omega}^t) \\
\text{s.t.} \ & \boldsymbol{x}_t \in X_t(\boldsymbol{x}^{t-1}, \boldsymbol{\omega}^t).
\end{aligned}$$

By assumption (A2)' and continuity of the expectation functional, the objective function of this maximization problem is continuous on Y^t. In addition, assumption (A3)' requires the feasible set mapping X_t to be continuous on Z^t. Thus, as in the basis step, Berge's theorem [13, theorem 12.1] guarantees continuity of Φ_t on Z^t. By induction, Φ_t is continuous for every $t \in \tau$, and $(E_t\Phi_{t+1})$ is continuous for every $t \in \tau_{-T}$. □

Although many stochastic programs of practical relevance comply with the regularity conditions (A1)–(A5), sometimes it might be necessary to circumvent or replace assumption (A4), which does not allow the distribution of the disturbances to depend on the outcome history. On these occasions, one may invoke a different qualification of the conditional probabilities to preserve upper semicontinuity of the expectation functionals. However, a much simpler approach is available in the presence of discrete probability measures. Then, upper semicontinuity of the value functions is not necessary to prove equivalence of the static and dynamic versions of a given stochastic program. In contrast, the following set of regularity conditions will suffice.

(A1)" = (A1); (A2)" = (A2); (A3)" = (A3);

(A4)" the probability measure $P_t(\cdot|\boldsymbol{\omega}^{t-1})$ is discrete with finite support

$$\operatorname{supp} P_t(\cdot|\boldsymbol{\omega}^{t-1}) = \left\{\omega_{t,i_t}(\boldsymbol{\omega}^{t-1}) \,\middle|\, i_t = 1, \ldots, I_t\right\},$$

and the probability of the i_t'th atom amounts to $p_{t,i_t}(\boldsymbol{\omega}^{t-1})$ for all $t \in \tau_{-0}$ and $\boldsymbol{\omega}^{t-1} \in \Omega^{t-1}$; in addition, the marginal measure P_0 has finite support with atoms ω_{0,i_0} and associated probabilities p_{0,i_0} for $i_0 = 1, \ldots, I_0$;

(A5)" = (A5).

By definition, P_t being a regular conditional probability requires the atoms ω_{t,i_t} and the associated probabilities p_{t,i_t} to be measurable functions on Ω^{t-1}, which will always be assumed implicitly. Notice that proposition 2.3 still holds under the conditions (A1)"–(A5)". Moreover, corollary 2.8 below is the counterpart of proposition 2.5.

Corollary 2.8. *Under the assumptions (A1)"–(A5)" the recourse function Φ_t is bounded, measurable in $\boldsymbol{\omega}^t$, and usc in $\boldsymbol{x}^{t-1}$ on its natural domain Z^t for all $t \in \tau_{-0}$. Moreover, Φ_0 is bounded and measurable on Ω_0.*

Proof. As usual, the claim is shown by backward induction with respect to t. Consider first the recourse function Φ_T. Boundedness and upper semicontinuity in $\boldsymbol{x}^{T-1}$ can be shown as in proposition 2.5, while measurability in $\boldsymbol{\omega}^T$ is straightforward. This observation completes the basis step.

Next, assume that the recourse function Φ_{t+1} is bounded, measurable in $\boldsymbol{\omega}^{t+1}$, and usc in $\boldsymbol{x}^t$ on Z^{t+1}. Consequently, the expectation functional

$$(E_t\Phi_{t+1})(\boldsymbol{x}^t, \boldsymbol{\omega}^t) = \sum_{i_t=1}^{I_t} \Phi_{t+1}(\boldsymbol{x}^t, \omega_{t+1,i_{t+1}}(\boldsymbol{\omega}^t), \boldsymbol{\omega}^t)\, p_{t+1,i_{t+1}}(\boldsymbol{\omega}^t)$$

is bounded, measurable in $\boldsymbol{\omega}^t$, and usc in $\boldsymbol{x}^t$ on Y^t by (A4)" and the induction hypothesis. Remember also that products and compositions of Borel measurable functions are Borel measurable. Next, consider the recourse function Φ_t. Boundedness and upper semicontinuity in $\boldsymbol{x}^{t-1}$ can be shown by arguing as in proposition 2.5, and measurability in $\boldsymbol{\omega}^t$ is obvious. These notions complete the induction step, and thus the claim follows. □

Corollary 2.9. *Under the assumptions (A1)"–(A5)" the static and dynamic versions of a multistage stochastic program are both solvable, and the optimal values coincide.*

Proof. Equivalence is straightforward since here integrals generally reduce to finite sums over the atoms of some marginal or conditional probability measure. Thus, interchanging the order of maximization and summation is unproblematic. Solvability is due to upper semicontinuity of the profit functions and the expectation functionals as well as compactness of the feasible sets. □

2.5 Here-and-Now Strategies

By definition, a strategy or policy is a decision rule $\tilde{\boldsymbol{x}}$ specifying the actions to be executed in each time step and in each scenario. Remember that non-anticipativity requires the measurable functions $\tilde{\boldsymbol{x}}_t$ to depend only on observations up to time t. In a more abstract formulation, the multistage stochastic program (2.8) reads

$$(E\Phi_0) = \sup_{\tilde{\boldsymbol{x}} \in \mathcal{N}_n} E(\hat{\rho}(\tilde{\boldsymbol{x}}(\tilde{\boldsymbol{\omega}}), \tilde{\boldsymbol{\omega}})). \tag{2.18}$$

Thereby, the objective function is given by the expectation value of

$$\hat{\rho}(\boldsymbol{x}, \boldsymbol{\omega}) := \sum_{t=0}^{T} \hat{\rho}_t(\boldsymbol{x}^t, \boldsymbol{\omega}^t).$$

Optimal policies $\tilde{\boldsymbol{x}}_{\mathrm{opt}} \in \arg\max_{\tilde{\boldsymbol{x}} \in \mathcal{N}_n} E(\hat{\rho}(\tilde{\boldsymbol{x}}(\tilde{\boldsymbol{\omega}}), \tilde{\boldsymbol{\omega}}))$ are called *here-and-now* strategies. On average, the implementation of a here-and-now strategy yields the optimal return $(E\Phi_0) = E(\hat{\rho}(\tilde{\boldsymbol{x}}_{\mathrm{opt}}(\tilde{\boldsymbol{\omega}}), \tilde{\boldsymbol{\omega}}))$.

2.6 Wait-and-See Strategies

If a decision maker can perfectly forecast the future outcomes of $\tilde{\boldsymbol{\omega}}$, he or she ends up with a deterministic *scenario problem* conditional on the anticipated realization $\boldsymbol{\omega}$:

$$\overline{\Phi}_0(\boldsymbol{\omega}) := \sup_{\boldsymbol{x} \in \mathbb{R}^n} \hat{\rho}(\boldsymbol{x}, \boldsymbol{\omega}) \tag{2.19}$$

Before the forecast, the expected earnings amount to $(E\overline{\Phi}_0) := E(\overline{\Phi}_0(\tilde{\boldsymbol{\omega}}))$. Notice that $(E\overline{\Phi}_0)$ would coincide with the result of (2.18) if the non-anticipativity restrictions were withdrawn. Thus, $(E\overline{\Phi}_0)$ is the result of the following optimization problem

$$(E\overline{\Phi}_0) = \sup_{\tilde{x} \in \mathcal{L}_n^\infty} E(\hat{\rho}(\tilde{x}(\tilde{\omega}), \tilde{\omega}))\,. \tag{2.20}$$

In fact, the mathematical program (2.20) has the same objective function as the here-and-now problem (2.18). However, the space of non-anticipative policies $\mathcal{N}_n$ is a subset of the space of anticipative policies $\mathcal{L}_n^\infty$, and thus $(E\overline{\Phi}_0)$ constitutes an upper bound for $(E\Phi_0)$. In particular, the indicator $\text{EVPI} := (E\overline{\Phi}_0) - (E\Phi_0)$ is a positive number and denotes the so-called *expected value of perfect information.* The EVPI was first introduced by Raiffa and Schlaifer [88], and it determines the value of knowing the realization of the random variable $\tilde{\omega}$ in advance, in contrast to knowing only the distribution of $\tilde{\omega}$. Alternatively, the EVPI can be interpreted as an insurance price or the maximum amount a rational actor would pay for perfect information about the future. Furthermore, optimal (anticipative) policies $\tilde{x}_{\text{ws}} \in \arg\max_{\tilde{x} \in \mathcal{L}_n^\infty} E(\hat{\rho}(\tilde{x}(\tilde{\omega}), \tilde{\omega}))$ are referred to as *wait-and-see* strategies. This terminology is due to Madansky [70]. As indicated above, a wait-and-see strategy achieves higher expected returns than a here-and-now strategy. However, its implementation requires perfect forecasts of the uncertain parameters; such forecasts are often expensive or unavailable.

2.7 Mean-Value Strategies

If a decision maker is unable or unwilling to take account of the stochasticity inherent to an optimization problem, he or she has to treat the random variables as deterministic parameters (whose values should be chosen with diligence). For instance, given a sequence of outcomes $\boldsymbol{\omega}^t$ known by observation, a reasonable approximation of the future uncertainties is given by their conditional expectation values[8] $E_t(\tilde{\omega}_s)$. Thus, at time t the decision maker under consideration solves the deterministic optimization problem

$$\sup_{\boldsymbol{x} \in \mathbb{R}^n} \left\{ \sum_{s=t}^{T} \hat{\rho}_s(\boldsymbol{x}^s, E_t(\tilde{\boldsymbol{\omega}}^s)) \,\middle|\, \boldsymbol{x}^{t-1} = \tilde{\boldsymbol{x}}_{\text{mv}}^{t-1}(\boldsymbol{\omega}^{t-1}) \right\}, \tag{2.21}$$

where the vector $\tilde{\boldsymbol{x}}_{\text{mv}}^{t-1}(\boldsymbol{\omega}^{t-1})$ contains those decisions, which have been implemented previously in response to the observations $\boldsymbol{\omega}^{t-1}$. Subsequently, the 'optimal' stage t decision $\tilde{\boldsymbol{x}}_{\text{mv},t}(\boldsymbol{\omega}^t)$ – as evaluated in (2.21) – is implemented, and the outcomes $\boldsymbol{\omega}_{t+1}$ are observed. Of course, the realization $\boldsymbol{\omega}_{t+1}$ of the random vector $\tilde{\boldsymbol{\omega}}_{t+1}$ is subject to the underlying conditional probability distribution and does not necessarily coincide with the conditional expectation $E_t(\tilde{\boldsymbol{\omega}}_{t+1})$ (as suggested in (2.21)). Therefore, the optimal stage $t+1$ decision of (2.21) is no longer adequate and could even be infeasible. In contrast, the

[8]Notice that the conditional expectation $E_t(\tilde{\boldsymbol{\omega}}_s)$ is an ordinary function of the outcomes $\boldsymbol{\omega}^t$ observed up to time t ($\forall s \in \tau$).

actions $\tilde{\boldsymbol{x}}_{\mathrm{mv},s}(\boldsymbol{\omega}^s)$ have to be determined with a rolling optimization scheme for $s = t+1, \ldots, T$. At every decision stage problem 2.21 must be recalibrated, i.e. the conditional expectations and the restrictions have to be updated consistently with the new observations and the implemented decisions.

By construction, the policy $\tilde{\boldsymbol{x}}_{\mathrm{mv}}$ obtained with this rolling optimization scheme is non-anticipative and feasible,[9] i.e. $\tilde{\boldsymbol{x}}_{\mathrm{mv}} \in \mathcal{N}_n$. The expected profit attained with this relatively modest decision strategy amounts to

$$(E\underline{\Phi}_0) := E(\hat{\rho}(\tilde{\boldsymbol{x}}_{\mathrm{mv}}(\tilde{\boldsymbol{\omega}}), \tilde{\boldsymbol{\omega}})). \tag{2.22}$$

Obviously, $(E\underline{\Phi}_0)$ is a lower bound for $(E\Phi_0) = E(\hat{\rho}(\tilde{\boldsymbol{x}}_{\mathrm{opt}}(\tilde{\boldsymbol{\omega}}), \tilde{\boldsymbol{\omega}}))$, since $\tilde{\boldsymbol{x}}_{\mathrm{opt}}$ maximizes the expected payoff over $\mathcal{N}_n$. Intuitively speaking, the error of disregarding the randomness of certain problem determinants lowers the achievable profit. Following Birge and Louveaux [10, Sect. 4.2], we can introduce the positive indicator VSS $:= (E\Phi_0) - (E\underline{\Phi}_0)$, which represents the *value of the stochastic solution* (alternative definitions are possible). The VSS measures the relative performance of the here-and-now strategy $\tilde{\boldsymbol{x}}_{\mathrm{opt}}$ as compared to the optimal mean-value strategy $\tilde{\boldsymbol{x}}_{\mathrm{mv}}$ and characterizes the maximum amount a rational decision maker would invest for incorporating uncertainty into an optimization problem.

The above reasoning remains valid for any deterministic process which is used to approximate the future uncertainties – taking the conditional expectation value is just an intuitive though arbitrary choice [8]. The numerical value of the VSS is certainly sensitive to the choice of the deterministic process, but it will always be positive.

---

[9] Using similar arguments as in the proof of theorem 2.6, it can be shown that $\tilde{\boldsymbol{x}}_{\mathrm{mv}}$ is a measurable mapping. Boundedness and non-anticipativity are straightforward.

3

Convex Stochastic Programs

In this chapter we focus on *convex* maximization problems. Thus, we assume the profit functions to be concave and the constraint functions to be convex in the decision variables. Under certain regularity conditions, which will be specified in Sect. 3.5, the recourse function of a convex multistage stochastic program is subdifferentiable and exhibits a characteristic saddle structure. We aim at exploiting these properties in order to discretize the underlying probability space in an efficient way.

Below, in order to simplify notation, random variables as well as their realizations will both be denoted by $\boldsymbol{\omega}_t$. Similarly, we use the same symbol $\boldsymbol{x}_t$ for the entire decision rule and the realized actions, as long as there is no risk of confusion.

3.1 Augmenting the Probability Space

The uncertain parameters $\boldsymbol{\omega}_t$ may influence both the profit and the constraint functions of any prototype optimization problem (2.8), which will be assumed to comply at least with the elementary regularity conditions (A1)–(A5). Based on this notion, Frauendorfer [42, Sect. 5] suggests a classification of the components of $\boldsymbol{\omega}_t$. Let us therefore introduce two subvectors $\boldsymbol{\eta}_t$ and $\boldsymbol{\xi}_t$ of the random vector $\boldsymbol{\omega}_t$, which are defined as follows:

(a) $\boldsymbol{\eta}_t$ comprises all components of $\boldsymbol{\omega}_t$ which influence the profit functions in a nontrivial way;

(b) $\boldsymbol{\xi}_t$ consists of those components of $\boldsymbol{\omega}_t$ which influence the constraint functions in a nontrivial way.

Some components of ω_t can be assigned exclusively either to η_t or ξ_t. In contrary, if this assignment is not unique for a specific component, we attach it to both subvectors and interpret it as a degenerate pair of uncertain parameters. The construction of η_t and ξ_t can be formalized by means of two problem-specific coordinate projections π_t^{o} and π_t^{r} $(t \in \tau)$.

$$\pi_t^{\mathrm{o}} : \begin{cases} \mathbb{R}^M \to \mathbb{R}^{K_t} \\ \omega \mapsto \pi_t^{\mathrm{o}}(\omega) = \eta_t \end{cases} \qquad \pi_t^{\mathrm{r}} : \begin{cases} \mathbb{R}^M \to \mathbb{R}^{L_t} \\ \omega \mapsto \pi_t^{\mathrm{r}}(\omega) = \xi_t \end{cases}$$

By assumption, $\Theta_t := \pi_t^{\mathrm{o}}(\Omega)$ is a compact subset of $\mathbb{R}^{K_t}$ and covers the support of η_t, whereas $\Xi_t := \pi_t^{\mathrm{r}}(\Omega)$ is a compact subset of $\mathbb{R}^{L_t}$ and covers the support of ξ_t. In complete analogy to the conventions met in Sect. 2.2, we form separate *outcome histories*

$$\begin{aligned} \eta^t &:= (\eta_0, \dots, \eta_t) \in \Theta^t := \Theta_0 \times \cdots \times \Theta_t \subset \mathbb{R}^{K^t}, \\ \xi^t &:= (\xi_0, \dots, \xi_t) \in \Xi^t := \Xi_0 \times \cdots \times \Xi_t \subset \mathbb{R}^{L^t}. \end{aligned}$$

As usual, the dimensions of the underlying Euclidean spaces are required to match up ($K^t = K_0 + \cdots + K_t$, $L^t = L_0 + \cdots + L_t$), and in the last decision stage the indices may be left out: $\eta := \eta^T$, $\Theta := \Theta^T \subset \mathbb{R}^K$, $\xi := \xi^T$, and $\Xi := \Xi^T \subset \mathbb{R}^L$ ($K := K^T$ and $L := L^T$). Formally speaking, the elements $\eta \in \Theta$ and $\xi \in \Xi$ corresponding to a specific outcome $\omega \in \Omega$ are obtained by means of the coordinate projections π^{o} and π^{r}.

$$\pi^{\mathrm{o}} := \pi_0^{\mathrm{o}} \times \cdots \times \pi_T^{\mathrm{o}} \qquad \pi^{\mathrm{r}} := \pi_0^{\mathrm{r}} \times \cdots \times \pi_T^{\mathrm{r}}$$

These projections enter the definition of a measurable transformation χ which relates Ω to $\Theta \times \Xi$ (cf. Fig. 3.1).

$$\chi : \begin{cases} \Omega \to \Theta \times \Xi \\ \omega \mapsto (\pi^{\mathrm{o}}(\omega), \pi^{\mathrm{r}}(\omega)) = (\eta, \xi) \end{cases}$$

The measurable space $(\Theta \times \Xi, \mathcal{B}(\Theta \times \Xi))$ basically inherits the probability measure P defined on $(\Omega, \mathcal{B}(\Omega))$. In fact, the induced probability measure P^{a} on $(\Theta \times \Xi, \mathcal{B}(\Theta \times \Xi))$ is characterized through

$$P^{\mathrm{a}}(A) := P(\chi^{-1}(A)) \quad \text{for} \quad A \in \mathcal{B}(\Theta \times \Xi).$$

In the sequel, the triple $(\Theta \times \Xi, \mathcal{B}(\Theta \times \Xi), P^{\mathrm{a}})$ is denoted as the *augmented probability space.* It can easily be verified that P^{a} is a regular probability measure, whose support is covered by $\Theta \times \Xi$. Notice that $P^{\mathrm{a}}(A) = 0$ if $A \cap \chi(\Omega) = \emptyset$. Next, define new *profit* and *constraint functions* on the augmented probability space.

$$\left.\begin{aligned} \rho_t^{\mathrm{a}}(x^t, \eta^t) &:= \rho_t(x^t, \omega^t) \\ f_t^{\mathrm{a}}(x^t, \xi^t) &:= f_t(x^t, \omega^t) \end{aligned}\right\} \quad \forall t \in \tau,\ (\eta, \xi) = \chi(\omega)$$

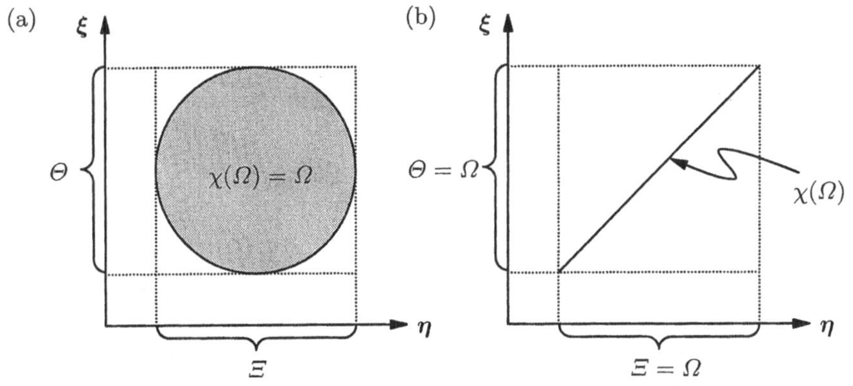

Fig. 3.1. **(a)** If there are no degenerate pairs of random variables, we find $\chi(\Omega) = \Omega$. Nevertheless, $\Theta \times \Xi$ does not necessarily coincide with Ω. Figure **(b)** shows the case of total degeneracy with $\Theta = \Xi = \Omega$. Generally speaking, degeneracy implies that $\chi(\Omega)$ has lower dimension than the product space $\Theta \times \Xi$

Observe that $\rho_t^{\mathrm{a}}(\boldsymbol{x}^t, \cdot)$ is a Borel measurable function on Θ^t instead of Ω^t, which is independent of $\boldsymbol{\xi}^t$. Similarly, $\boldsymbol{f}_t^{\mathrm{a}}(\boldsymbol{x}^t, \cdot)$ represents a Borel measurable mapping on Ξ^t being independent of $\boldsymbol{\eta}^t$. In fact, Borel measurability of ρ_t^{a} and $\boldsymbol{f}_t^{\mathrm{a}}$ follows from Borel measurability of ρ_t and $\boldsymbol{f}_t$, respectively, as may be seen from part (1) of the proof of [2, theorem 2.6.4]. Then, let us introduce the *static version* of a stochastic program on the augmented probability space.

$$\begin{aligned} &\sup_{\boldsymbol{x} \in \mathcal{N}_n^{\mathrm{a}}} \int_{\Theta \times \Xi} \left[\sum_{t=0}^{T} \rho_t^{\mathrm{a}}(\boldsymbol{x}^t, \boldsymbol{\eta}^t) \right] dP^{\mathrm{a}}(\boldsymbol{\eta}, \boldsymbol{\xi}) \\ &\text{s.t.} \quad \boldsymbol{f}_t^{\mathrm{a}}(\boldsymbol{x}^t, \boldsymbol{\xi}^t) \leq \mathbf{0} \quad P^{\mathrm{a}}\text{-a.s.} \quad t \in \tau \end{aligned} \tag{3.1}$$

Here, $\mathcal{N}_n^{\mathrm{a}}$ represents the space of essentially bounded non-anticipative policy functions on $(\Theta \times \Xi, \mathcal{B}(\Theta \times \Xi), P^{\mathrm{a}})$. By means of the change of variable formula of integration theory it can be shown that (3.1) is equivalent to the underlying original problem (2.8), see e.g. [42, Sect. 5]. In fact, both problems have the same optimal values, and their optimizers are related by the measurable transformation χ.

Intuitively speaking, there is no loss of information if we ignore those subspaces on which the profit and constraint functions are constant. Thus, ρ_t^{a} and $\boldsymbol{f}_t^{\mathrm{a}}$ will be identified with ρ_t and $\boldsymbol{f}_t$ in the remainder, and the superscript 'a' is suppressed to simplify notation. Moreover, for the sake of transparency we will write P and $\mathcal{N}_n$ instead of P^{a} and $\mathcal{N}_n^{\mathrm{a}}$, respectively. Sometimes it is useful to bring the explicit constraints to the objective function. Therefore, in complete analogy to (2.9), we introduce the *effective profit functions*

$$\hat{\rho}_t(\boldsymbol{x}^t, \boldsymbol{\eta}^t, \boldsymbol{\xi}^t) := \begin{cases} \rho_t(\boldsymbol{x}^t, \boldsymbol{\eta}^t) & \text{for } \boldsymbol{f}_t(\boldsymbol{x}^t, \boldsymbol{\xi}^t) \leq \mathbf{0}, \\ -\infty & \text{else.} \end{cases}$$

Based on the above definitions in connection with the augmented probability space, we can also state the *dynamic version* of the stochastic program (3.1). First of all, the parametric family of stage T subproblems reads

$$\Phi_T(\boldsymbol{x}^{T-1},\boldsymbol{\eta}^T,\boldsymbol{\xi}^T) := \sup_{\boldsymbol{x}_T\in\mathbb{R}^{n_T}} \hat{\rho}_T(\boldsymbol{x}^T,\boldsymbol{\eta}^T,\boldsymbol{\xi}^T), \tag{3.2a}$$

whereas the subproblems for $t = 1,\ldots,T-1$ are given by

$$\Phi_t(\boldsymbol{x}^{t-1},\boldsymbol{\eta}^t,\boldsymbol{\xi}^t) := \sup_{\boldsymbol{x}_t\in\mathbb{R}^{n_t}} \hat{\rho}_t(\boldsymbol{x}^t,\boldsymbol{\eta}^t,\boldsymbol{\xi}^t) + (E_t\Phi_{t+1})(\boldsymbol{x}^t,\boldsymbol{\eta}^t,\boldsymbol{\xi}^t). \tag{3.2b}$$

On the other hand, the zero stage problems can be written as

$$\Phi_t(\boldsymbol{\eta}_0,\boldsymbol{\xi}_0) := \sup_{\boldsymbol{x}_0\in\mathbb{R}^{n_0}} \hat{\rho}_t(\boldsymbol{x}_0,\boldsymbol{\eta}_0,\boldsymbol{\xi}_0) + (E_0\Phi_1)(\boldsymbol{x}_0,\boldsymbol{\eta}_0,\boldsymbol{\xi}_0). \tag{3.2c}$$

The optimal value functions in (3.2) are usually called *recourse functions.* Moreover, we made use of the *expectation functionals* $(E_t\Phi_{t+1})$, which are defined in the obvious way for $t \in \tau_{-T}$.

$$(E_t\Phi_{t+1})(\boldsymbol{x}^t,\boldsymbol{\eta}^t,\boldsymbol{\xi}^t) := \int \Phi_{t+1}(\boldsymbol{x}^t,\boldsymbol{\eta}^{t+1},\boldsymbol{\xi}^{t+1})dP_{t+1}(\boldsymbol{\eta}_{t+1},\boldsymbol{\xi}_{t+1}|\boldsymbol{\eta}^t,\boldsymbol{\xi}^t) \tag{3.3}$$

The optimal value $(E\Phi_0) := \int \Phi_0 dP_0$ corresponding to the dynamic program (3.2) is given by the unconditional expectation of the recourse function Φ_0. The *feasible set mappings*

$$\boldsymbol{\xi}_0 \mapsto X_0(\boldsymbol{\xi}_0) \quad \text{and} \quad (\boldsymbol{x}^{t-1},\boldsymbol{\xi}^t) \mapsto X_t(\boldsymbol{x}^{t-1},\boldsymbol{\xi}^t) \quad \text{for } t \in \tau_{-0}$$

are defined as in Sect. 2.3 and can be interpreted as multifunctions which are independent of $\boldsymbol{\eta}$. Then, for any $t \in \tau$ the *nested feasible set mapping* X^t is defined through

$$X^t : \begin{cases} \Xi^t \to 2^{\mathbb{R}^{n^t}} \\ \boldsymbol{\xi}^t \mapsto \left\{\boldsymbol{x}^t \in \mathbb{R}^{n^t} \,\middle|\, \boldsymbol{x}_0 \in X_0(\boldsymbol{\xi}_0),\ldots,\boldsymbol{x}_t \in X_t(\boldsymbol{x}^{t-1},\boldsymbol{\xi}^t)\right\}. \end{cases}$$

Many statements about the dynamic program (3.2) are most conveniently expressed in terms of the *generalized feasible sets* Y^t, which comprise the joint outcome and decision histories feasible up to time t, $t \in \tau$. Note that, unlike in Chap. 2, Y^t is now built on the augmented probability space. For every $t \in \tau$ set

$$Y^t := \left\{(\boldsymbol{x}^t,\boldsymbol{\eta}^t,\boldsymbol{\xi}^t)\middle|\boldsymbol{\eta}^t \in \Theta^t,\ \boldsymbol{\xi}^t \in \Xi^t,\ \boldsymbol{x}^t \in X^t(\boldsymbol{\xi}^t)\right\}.$$

In fact, Y^t is interpreted as the *natural domain* of the profit function ρ_t, the expectation functional $(E_t\Phi_{t+1})$, and the constraint function $\boldsymbol{f}_t$. For the sake of transparent notation we need the related *generalized feasible sets* Z^t, which contain no stage t decisions, $t \in \tau$. Set $Z^0 := \Theta_0 \times \Xi_0$, and for $t \in \tau_{-0}$ define

$$Z^t := \left\{(\boldsymbol{x}^{t-1},\boldsymbol{\eta}^t,\boldsymbol{\xi}^t)\middle|\boldsymbol{\eta}^t \in \Theta^t,\ \boldsymbol{\xi}^t \in \Xi^t,\ \boldsymbol{x}^{t-1} \in X^{t-1}(\boldsymbol{\xi}^{t-1})\right\}.$$

Notice that Z^t basically constitutes the *natural domain* of the recourse function of stage t.

3.2 Preliminary Definitions

Once having introduced the basic notation and terminology, we should turn to the formulation of appropriate regularity conditions for problem (3.1), which are slightly stronger than (A1)–(A5). For instance, we are in need of suitable *constraint qualifications* (see e.g. [6, Chap. 5] for a survey of common constraint qualifications and their use for validating optimality conditions). Thus, in Sect. 3.2.1 we will discuss *Slater's constraint qualification*, always taking account of the fact that $\boldsymbol{f}_t := (\boldsymbol{f}_t^{\text{in}}, \boldsymbol{f}_t^{\text{eq}}, -\boldsymbol{f}_t^{\text{eq}})$ includes both inequality and equality constraints. Furthermore, in Sect. 3.2.2 we will introduce a generalized notion of convexity allowing for extended-real-valued functions defined on nonconvex domains. Finally, Sect. 3.2.3 is devoted to the study of a class of autoregressive stochastic processes, which will play a major role in proving the main results of the present chapter.

3.2.1 Slater's Constraint Qualification

In this section we introduce Slater's constraint qualification and demonstrate its importance in the field of parametric optimization. For the sake of transparent notation, we consider a simple parametric optimization problem over the Euclidean space $\mathbb{R}^n$,

$$\begin{aligned} &\sup_{\boldsymbol{x}\in\mathbb{R}^n} \rho(\boldsymbol{x},\boldsymbol{\omega}) \\ \text{s.t.}\quad &\boldsymbol{f}^{\text{in}}(\boldsymbol{x},\boldsymbol{\omega}) \leq \mathbf{0} \\ &\boldsymbol{f}^{\text{eq}}(\boldsymbol{x},\boldsymbol{\omega}) = \mathbf{0} \end{aligned} \tag{3.4}$$

and the parameter $\boldsymbol{\omega}$ ranges over $\mathbb{R}^M$. The extended-real-valued objective function $\rho : \mathbb{R}^n \times \mathbb{R}^M \to [-\infty, \infty]$ is assumed to be upper semicontinuous, whereas the vector-valued constraint functions

$$\begin{aligned} \boldsymbol{f}^{\text{in}} &: \mathbb{R}^n \times \mathbb{R}^M \to \mathbb{R}^{r^{\text{in}}} \\ \boldsymbol{f}^{\text{eq}} &: \mathbb{R}^n \times \mathbb{R}^M \to \mathbb{R}^{r^{\text{eq}}} \end{aligned}$$

are continuous. As usual, the corresponding feasible set mapping

$$X : \mathbb{R}^M \to 2^{\mathbb{R}^n}, \quad \boldsymbol{\omega} \mapsto X(\boldsymbol{\omega}) := \{\boldsymbol{x} \,|\, \boldsymbol{f}^{\text{in}}(\boldsymbol{x},\boldsymbol{\omega}) \leq \mathbf{0},\ \boldsymbol{f}^{\text{eq}}(\boldsymbol{x},\boldsymbol{\omega}) = \mathbf{0}\}$$

characterizes a closed-valued multifunction. Furthermore, for some reference point $\bar{\boldsymbol{\omega}} \in \mathbb{R}^M$, the equality constraints are of the form

$$f_i^{\text{eq}}(\boldsymbol{x},\boldsymbol{\omega}) \equiv \langle \boldsymbol{w}_i, \boldsymbol{x} \rangle - h_i(\boldsymbol{\omega}) \quad \forall i = 1, \ldots, r^{\text{eq}} \tag{3.5}$$

on an open neighborhood V of $\bar{\boldsymbol{\omega}}$, where the gradient $\boldsymbol{w}_i \in \mathbb{R}^n$ is a constant vector[1] independent of $\boldsymbol{\omega}$, and $h_i : V \to \mathbb{R}$ is a continuous functional for each $i = 1, \ldots, r^{\text{eq}}$.

[1] Allowing for constraint functions of type (3.5), whose gradients are not constant on the entire space, will avoid tedious case differentiations in Chap. 5.

Definition 3.1. *The parametric maximization problem (3.4) satisfies Slater's constraint qualification at the reference point* $\bar{\boldsymbol{\omega}}$ *if there is a decision vector* $\boldsymbol{y} \in \mathbb{R}^n$ *such that*

$$\boldsymbol{f}^{\mathrm{in}}(\boldsymbol{y}, \bar{\boldsymbol{\omega}}) < \boldsymbol{0} \quad \text{and} \quad \boldsymbol{f}^{\mathrm{eq}}(\boldsymbol{y}, \bar{\boldsymbol{\omega}}) = \boldsymbol{0},$$

and the gradients of the equality constraints (3.5) are linearly independent. The decision $\boldsymbol{y}$ *is referred to as a Slater point.*

Linear independence of the gradients $\boldsymbol{w}_i$, $i = 1, \ldots, r^{\mathrm{eq}}$, automatically holds true if redundant constraints are deleted. Therefore, this requirement is no real restriction. Slater's constraint qualification is extremely useful when dealing with parametric optimization problems since it preserves feasibility. Concretely speaking, assume that Slater's constraint qualification holds at the given reference point $\bar{\boldsymbol{\omega}}$. Then, (3.4) is not only feasible at the reference point but also in its vicinity (see proposition 3.2 below).

Proposition 3.2. *If the parametric optimization problem (3.4) fulfills Slater's constraint qualification at the reference point* $\bar{\boldsymbol{\omega}} \in \mathbb{R}^M$*, then the multifunction* X *has a continuous selector on a neighborhood* V *of* $\bar{\boldsymbol{\omega}}$*. Formally speaking, there exists a continuous function*

$$\tilde{\boldsymbol{x}} : V \to \mathbb{R}^n \quad \text{such that} \quad \tilde{\boldsymbol{x}}(\boldsymbol{\omega}) \in X(\boldsymbol{\omega}) \quad \forall \boldsymbol{\omega} \in V.$$

Proof. Let $\boldsymbol{y}$ be a Slater point in $X(\bar{\boldsymbol{\omega}})$, which is kept fixed. By continuity of the constraint functions there is a neighborhood $U \times V'$ of $(\boldsymbol{y}, \bar{\boldsymbol{\omega}})$ such that

$$\boldsymbol{f}^{\mathrm{in}}(\boldsymbol{x}, \boldsymbol{\omega}) < \boldsymbol{0} \quad \forall (\boldsymbol{x}, \boldsymbol{\omega}) \in U \times V'.$$

If there are no equality constraints, we may simply set $V := V'$ and $\tilde{\boldsymbol{x}}(\boldsymbol{\omega}) \equiv \boldsymbol{y}$, and the claim follows trivially. In the presence of equality constraints, however, some additional work is necessary. By assumption, the equality constraints are of the form (3.5) on an open neighborhood V'' of $\bar{\boldsymbol{\omega}}$, and the vectors $\boldsymbol{w}_i$, $i = 1, \ldots, r^{\mathrm{eq}}$, are linearly independent. Moreover, the real-valued functionals h_i, $i = 1, \ldots, r^{\mathrm{eq}}$ are continuous on V''. If $r^{\mathrm{eq}} < n$, we may complement the given constraint set by additional fictitious constraints of the form

$$f_i^{\mathrm{eq}}(\boldsymbol{x}, \boldsymbol{\omega}) \equiv \langle \boldsymbol{w}_i, \boldsymbol{x} \rangle - h_i(\boldsymbol{\omega}) \quad \forall i = r^{\mathrm{eq}} + 1, \ldots, n,$$

where $h_i(\boldsymbol{\omega}) := \langle \boldsymbol{w}_i, \boldsymbol{y} \rangle$ is constant. The additional equality constraints are fully determined by a family of new vectors $\boldsymbol{w}_i$, $i = r^{\mathrm{eq}} + 1, \ldots, n$. These vectors are chosen such that $\{\boldsymbol{w}_i\}_{i=1}^n$ is a basis of $\mathbb{R}^n$. For the further argumentation we need an $n \times n$ matrix W with row vectors $\boldsymbol{w}_1, \ldots, \boldsymbol{w}_n$. In addition, we combine the functionals h_i to a continuous vector-valued mapping $\boldsymbol{h} := (h_1, \ldots, h_n)$ on V''. By construction, W is regular and can be inverted. Thus, we may introduce a well-defined function

$$\tilde{\boldsymbol{x}}(\boldsymbol{\omega}) := W^{-1}\boldsymbol{h}(\boldsymbol{\omega}).$$

Obviously, $\tilde{\boldsymbol{x}}$ is continuous on V'', and we have $\tilde{\boldsymbol{x}}(\bar{\boldsymbol{\omega}}) = \boldsymbol{y}$. Thus, there is a neighborhood V of $\bar{\boldsymbol{\omega}}$ such that $V \subset V' \cap V''$ and

$$\tilde{\boldsymbol{x}}(\boldsymbol{\omega}) \in U \quad \forall(\boldsymbol{x}, \boldsymbol{\omega}) \in V.$$

Finally, one may check that the decision $\tilde{\boldsymbol{x}}(\boldsymbol{\omega})$ satisfies the constraints

$$\boldsymbol{f}^{\mathrm{eq}}(\tilde{\boldsymbol{x}}(\boldsymbol{\omega}), \boldsymbol{\omega}) = \boldsymbol{0} \quad \text{and} \quad \boldsymbol{f}^{\mathrm{in}}(\tilde{\boldsymbol{x}}(\boldsymbol{\omega}), \boldsymbol{\omega}) < \boldsymbol{0}$$

for all parameters $\boldsymbol{\omega} \in V$. This observation completes the proof. □

Corollary 3.3. *Under the assumptions of proposition 3.2 the parametric optimization problem (3.4) satisfies Slater's constraint qualification at every* $\boldsymbol{\omega}$ *in some neighborhood of* $\bar{\boldsymbol{\omega}}$.

Proof. Consider the mapping $\tilde{\boldsymbol{x}} : V \to \mathbb{R}^n$ in the proof of proposition 3.2. For any $\boldsymbol{\omega} \in V$ one can easily check that $\tilde{\boldsymbol{x}}(\boldsymbol{\omega})$ represents a Slater point for the optimization problem (3.4) with respect to the reference point $\boldsymbol{\omega}$. □

3.2.2 Convex Functions

The formulation of appropriate regularity conditions for problem (3.1) not only requires suitable constraint qualifications but also a precise definition of *convex* and *concave functions*. Classical literature on convex analysis usually assumes the domain of convex functions to be convex [89]. However, in the remainder of this work we will sometimes need a generalized notion of convexity allowing for extended-real-valued functions defined on nonconvex sets. Such functions will naturally arise in Chap. 5. Moreover, for the sake of transparent terminology, we aim at extending the notion of convexity to vector-valued mappings, as is done e.g. in [67]. This will be particularly useful when dealing with constraint functions. The following definition provides a precise characterization of convexity as understood in this work.

Definition 3.4. (Convex Functions)

(i) A function $\rho : \mathbb{R}^n \to [-\infty, \infty]$ *is convex on an arbitrary set* $U \subset \mathbb{R}^n$ *if for every choice of* $\boldsymbol{x}_0 \in U$ *and* $\boldsymbol{x}_1 \in U$ *one has*

$$\rho(\boldsymbol{x}_\lambda) \leq (1-\lambda)\rho(\boldsymbol{x}_0) + \lambda\rho(\boldsymbol{x}_1)$$

for all $\lambda \in (0, 1)$ *such that* $\boldsymbol{x}_\lambda = (1-\lambda)\boldsymbol{x}_0 + \lambda\boldsymbol{x}_1 \in U$.

(ii) A mapping $\boldsymbol{f} : \mathbb{R}^n \to [-\infty, \infty]^r$ *is convex on an arbitrary set* $U \subset \mathbb{R}^n$ *if each component* f_i *is convex on* U *for* $i = 1, \ldots, r$.

Infinite values of convex functions are handled by using *extended arithmetic* as explained in [97, Sect. 1.E]. Most importantly, one should use the convention $\infty - \infty = -\infty$. An extended-real-valued function ρ is said to be concave if $-\rho$ is convex; similarly, a mapping $\boldsymbol{f}$ is concave if $-\boldsymbol{f}$ is convex. When dealing with concave functions, one usually adopts the convention $\infty - \infty = \infty$, which gives rise to a slightly different form of extended arithmetic.

3.2.3 Block-Diagonal Autoregressive Processes

In the present chapter we are interested in stochastic programs whose recourse functions can be shown to exhibit a specific saddle structure. This forces us to restrict attention to a specific subclass of the nonlinear autoregressive processes considered in Chap. 2. Concretely speaking, we will have to study the class of so-called block-diagonal autoregressive processes.

Definition 3.5. *We say that the random data $\{\boldsymbol{\eta}_t, \boldsymbol{\xi}_t\}_{t\in\mathcal{T}}$ follows a block-diagonal autoregressive process if for every $t \in \mathcal{T}_{-0}$ there are two possibly correlated random vectors $(\boldsymbol{\varepsilon}_t^o, \boldsymbol{\varepsilon}_t^r)$ with induced probability space $(\mathcal{E}_t^o \times \mathcal{E}_t^r, \mathcal{B}(\mathcal{E}_t^o \times \mathcal{E}_t^r), Q_t)$ and two matrices $H_t^o \in \mathbb{R}^{K_t \times K^{t-1}}$ and $H_t^r \in \mathbb{R}^{L_t \times L^{t-1}}$ such that*

$$\begin{bmatrix} \boldsymbol{\eta}_t \\ \boldsymbol{\xi}_t \end{bmatrix} = \begin{bmatrix} H_t^o & 0 \\ 0 & H_t^r \end{bmatrix} \begin{bmatrix} \boldsymbol{\eta}^{t-1} \\ \boldsymbol{\xi}^{t-1} \end{bmatrix} + \begin{bmatrix} \boldsymbol{\varepsilon}_t^o \\ \boldsymbol{\varepsilon}_t^r \end{bmatrix}.$$

The initial data $(\boldsymbol{\eta}_0, \boldsymbol{\xi}_0)$ and the disturbances $\{\boldsymbol{\varepsilon}_t^o, \boldsymbol{\varepsilon}_t^r\}_{t\in\mathcal{T}_{-0}}$ are mutually independent, and for each $t \in \mathcal{T}_{-0}$ we postulate:

- *$\mathcal{E}_t^o \subset \mathbb{R}^{K_t}$ and $\mathcal{E}_t^r \subset \mathbb{R}^{L_t}$ are Borel sets;*
- *Q_t is a regular probability measure on $(\mathcal{E}_t^o \times \mathcal{E}_t^r, \mathcal{B}(\mathcal{E}_t^o \times \mathcal{E}_t^r))$.*

Block-diagonal autoregressive processes exhibit a linear dependence on the outcome history. Moreover, the AR coefficient matrices of all stages are block-diagonal. The reason for this nonstandard requirement will become clear in Sect. 3.5.

3.3 Regularity Conditions

Using the notation introduced so far, we require the stochastic program (3.1) on the augmented probability space to satisfy the conditions:

(B1) – Θ_t is a compact simplex in $\mathbb{R}^{K_t}$ covering the support of the random vector $\boldsymbol{\eta}_t$, $t \in \mathcal{T}$;
– Ξ_t is a compact simplex in $\mathbb{R}^{L_t}$ covering the support of the random vector $\boldsymbol{\xi}_t$, $t \in \mathcal{T}$;

(B2) – the profit function ρ_t is continuous on the underlying Euclidean space and concave in $\boldsymbol{x}^t$ for fixed values of the stochastic parameters, $t \in \mathcal{T}$;
– there is a convex neighborhood of Y^t on which ρ_t is a saddle function being concave in $\boldsymbol{x}^t$, convex in $\boldsymbol{\eta}^t$, and constant in $\boldsymbol{\xi}^t$, $t \in \mathcal{T}$;

(B3) – the constraint function $\boldsymbol{f}_t$ is continuous on the underlying Euclidean space and convex in $\boldsymbol{x}^t$ for fixed values of the stochastic parameters, $t \in \mathcal{T}$;
– there is a convex neighborhood of Y^t on which $\boldsymbol{f}_t$ is jointly convex in $(\boldsymbol{x}^t, \boldsymbol{\xi}^t)$ and constant in $\boldsymbol{\eta}^t$, $t \in \mathcal{T}$;
– the feasible set mapping X_t is bounded on a neighborhood of Z^t, $t \in \mathcal{T}$;

(B4) the random data $\{\boldsymbol{\eta}_t, \boldsymbol{\xi}_t\}_{t\in\mathcal{T}}$ follows a block-diagonal autoregressive process; $\mathcal{E}_t^{\mathrm{o}}$ and $\mathcal{E}_t^{\mathrm{r}}$ are compact simplices for all $t \in \mathcal{T}_{-0}$;

(B5) the parametric maximization problems (3.2) fulfill Slater's constraint qualification at any reference point in Z^t, $t \in \mathcal{T}$.

Assumption (B1) implies that the marginal spaces Θ_t and Ξ_t are simplicial sets with non-empty interior. This, of course, can always be enforced by appropriate definitions, given that the support of the random variables is bounded. As argued in Sect. 2.4, the solution of a stochastic program depends only on the restriction of ρ_t and $\boldsymbol{f}_t$ to the natural domain Y^t for every $t \in \mathcal{T}$. Thus, one might expect that (B2) and (B3) should exclusively qualify the behavior of ρ_t and $\boldsymbol{f}_t$ on Y^t. However, specific structural properties of the profit and constraint functions are required to hold on a neighborhood of Y^t. This generalization is necessary for technical reasons and allows us to prove subdifferentiability of the recourse functions below. Assumption (B3) implies that the feasible set mapping X_t is convex-closed-valued. Moreover, (B3) requires both $\boldsymbol{f}_t^{\mathrm{eq}}$ and $-\boldsymbol{f}_t^{\mathrm{eq}}$ to be convex in their arguments implying that all equality constraints are linear affine on a convex neighborhood of Y^t. Assumption (B4) basically states that random data follows a block-diagonal autoregressive process driven by serially uncorrelated noise. The matrices H_t^{o} and H_t^{r} contain the non-vanishing AR coefficients. Finally, condition (B5) states that the values of the feasible set mapping X_t have non-empty relative interior on the natural domain Z^t due to the existence of Slater points and continuity of the constraint functions. This implies non-anticipativity of the constraint multifunction.

In the remainder of this section we will derive some elementary implications of the new regularity conditions. In particular, we will argue that (B1)–(B5) entail the more fundamental conditions (A1)–(A5).

Proposition 3.6. *Under the assumptions (B1) and (B3) the generalized feasible set Y^t is convex for $t \in \tau$.*

Proof. The assertion follows immediately from the convexity properties of the constraint functions. □

Proposition 3.7. *Under the assumptions (B1)–(B5) the feasible set mapping X_t is non-empty-valued on a neighborhood of Z^t, $t \in \tau$.*

Proof. Apply proposition 3.2 to the parametric maximization problems (3.2a), (3.2b), and (3.2c) at an arbitrary reference point in Z^t, $t \in \tau$. Consequently, there exists a feasible decision vector for every outcome and decision history in a neighborhood of the reference point. As the reference point can be chosen freely in Z^t, the claim follows. □

Proposition 3.8. *Under the assumptions (B1)–(B5) the multifunction X_t is upper semicontinuous on a neighborhood of Z^t for all $t \in \tau$.*

Proof. By boundedness of the feasible set mapping X_t, there is a neighborhood V of Z^t such that $X_t(V)$ represents a bounded subset of $\mathbb{R}^{n_t}$. Moreover, continuity of the constraint functions implies that the graph of the multifunction X_t is closed. These observations allow us to invoke [13, proposition 11.9 (b)] to prove that X_t is usc on V. □

Proposition 3.9. *Under the regularity conditions (B1)–(B5) the multifunction X_t is compact-valued on a neighborhood of Z^t, $t \in \tau$.*

Proof. This is immediate from assumption (B3). □

Proposition 3.10. *Consider a stochastic program subject to the regularity conditions (B1)–(B5). Then, for any $t \in \tau$ and for every neighborhood U of Y^t there is a neighborhood V of Z^t such that the graph of X_t over V is a subset of U, i.e.*

$$\operatorname{graph} X_t|_V \subset U.$$

Proof. Without loss of generality we may consider the case $t > 0$. Select a reference point $(\bar{\boldsymbol{x}}^{t-1}, \bar{\boldsymbol{\eta}}^t, \bar{\boldsymbol{\xi}}^t) \in Z^t$. Since the feasible set mapping X_t is compact-valued on Z^t, there exist open sets V' and W with

$$(\bar{\boldsymbol{x}}^{t-1}, \bar{\boldsymbol{\eta}}^t, \bar{\boldsymbol{\xi}}^t) \in V', \quad X_t(\bar{\boldsymbol{x}}^{t-1}, \bar{\boldsymbol{\xi}}^t) \subset W, \quad \text{and} \quad V' \times W \subset U.$$

Furthermore, since X_t is usc on Z^t, there is an open neighborhood V'' of the reference point such that

$$X_t(\boldsymbol{x}^{t-1}, \boldsymbol{\xi}^t) \subset W \quad \forall (\boldsymbol{x}^{t-1}, \boldsymbol{\eta}^t, \boldsymbol{\xi}^t) \in V''.$$

Therefore, the graph of X_t over the open set $V' \cap V''$ is covered by U. As the choice of the reference point was arbitrary, there is a neighborhood V of Z^t such that the graph of X_t over V is contained in U. □

We may use the above results to show that the assumptions (B1)–(B5) imply the fundamental regularity conditions (A1)–(A5). Consider thus a stochastic program of the form (3.1) subject to (B1)–(B5). Then, the conditions (A1), (A4), and (A5) are trivially fulfilled.[2] Moreover, proposition 3.8 entails (A3), which is equivalent to compactness of the generalized feasible sets $\{Y_t\}_{t \in \tau}$, as implied by proposition 2.3. Assumption (B2) requires the profit function ρ_t to be continuous (and a fortiori usc) on a convex neighborhood of Y^t for all $t \in \tau$. Furthermore, the Weierstrass maximum theorem ensures that ρ_t is bounded on Y^t. Thus, (A2) follows. In summary, the stochastic program (3.1) satisfies each of the fundamental regularity conditions (A1)–(A5). Therefore, the statements of proposition 2.5 and theorem 2.6 hold, i.e. the static and dynamic versions of the stochastic program (3.1) are well-defined and solvable, and the optimal values coincide.

Proposition 3.8 proves upper semicontinuity of the feasible set mapping X_t. This is not the sharpest result available. In addition, as X_t has a convex graph, it is easily shown to be lower semicontinuous on a neighborhood of Z^t. This follows from theorem 5.9 (b) in [97]. Notice, however, that lower semicontinuity is referred to as 'inner semicontinuity' in that reference.

Being both usc and lsc, the feasible set mapping X_t is continuous on its natural domain Z^t. This observation allows us to prove that the assumptions (B1)–(B5) also imply (A1)'–(A5)'. Therefore, given the specific regularity conditions (B1)–(B5), the recourse functions are continuous on their natural domains by virtue of proposition 2.7. However, below we will argue that the recourse functions are not only continuous but also subdifferentiable and saddle-shaped. These properties will turn out to be more important than continuity.

[2] In order to check these regularity conditions set $\Omega := \Theta \times \Xi$ and $\omega := (\eta, \xi)$.

3.4 sup-Projections

In order to infer the saddle property of the recourse functions, we need two basic results about sup-projections, which are due to Rockafellar (see [89, theorem 5.5] and [90, theorem 1]).

Proposition 3.11. *Let $\rho(\boldsymbol{x},\boldsymbol{\eta})$ be an extended-real-valued function on $\mathbb{R}^n \times \mathbb{R}^K$ which is convex in $\boldsymbol{\eta}$. Then, the sup-projection $\Phi(\boldsymbol{\eta}) := \sup_{\boldsymbol{x}} \rho(\boldsymbol{x},\boldsymbol{\eta})$ is a convex function on $\mathbb{R}^K$.*

Proof. Consider the family $\{\rho(\boldsymbol{x},\cdot)\}_{\boldsymbol{x}\in\mathbb{R}^n}$ of convex functions indexed by the parameter $\boldsymbol{x}$. The epigraph of Φ is characterized by the infinite intersection of the convex epigraphs of $\rho(\boldsymbol{x},\cdot)$, $\boldsymbol{x} \in \mathbb{R}^n$, and constitutes a convex set. Consequently, Φ is a convex function on $\mathbb{R}^K$. □

Proposition 3.12. *Let $\rho(\boldsymbol{x},\boldsymbol{\xi})$ be an extended-real-valued function on $\mathbb{R}^n \times \mathbb{R}^L$ which is jointly concave in $\boldsymbol{x}$ and $\boldsymbol{\xi}$. Then, the sup-projection $\Phi(\boldsymbol{\xi}) := \sup_{\boldsymbol{x}} \rho(\boldsymbol{x},\boldsymbol{\xi})$ is a concave function on $\mathbb{R}^L$.*

Proof. Denote by E the projection of $\operatorname{hypo}\rho$ on $\mathbb{R}^L \times \mathbb{R}$. The hypograph of Φ coincides with E except for some special boundary points (the 'vertical' fibres of $\operatorname{hypo}\Phi$ must be closed for all fixed vectors $\boldsymbol{\xi} \in \mathbb{R}^L$):

$$\operatorname{hypo}\Phi = \{(\boldsymbol{\xi},\alpha) \in \mathbb{R}^L \times \mathbb{R} \,|\, (\boldsymbol{\xi},\beta) \in E \quad \forall \beta < \alpha\}.$$

By assumption, the hypograph of the objective function ρ is convex, and since convexity is preserved under projections, E is convex, as well. This notion entails convexity of $\operatorname{hypo}\Phi$, and therefore Φ is a concave function on $\mathbb{R}^L$. □

The conditions of proposition 3.12 can not be relaxed to allow for *biconcave* functions ρ. For example, consider the following extended-real-valued mapping defined on $\mathbb{R}^2$:

$$\rho(x,\xi) := \begin{cases} x\xi & \text{for } (x,\xi) \in [-1,1] \times \mathbb{R}, \\ -\infty & \text{else.} \end{cases}$$

Apparently, $\rho(\cdot,\xi)$ is concave in x for every fixed value of ξ, and the parametric function $\rho(x,\cdot)$ is concave in ξ. However, ρ is *not* jointly concave in x and ξ. The corresponding sup-projection

$$\Phi(\xi) := \sup_x \rho(x,\xi) = |\xi| \quad \forall \xi \in \mathbb{R}$$

is convex instead of being concave. Figure 3.2 depicts the nonconvex hypograph of ρ and demonstrates the construction of $\operatorname{hypo}\Phi$. Thus, sup-projections of biconcave functions ρ are not necessarily concave.

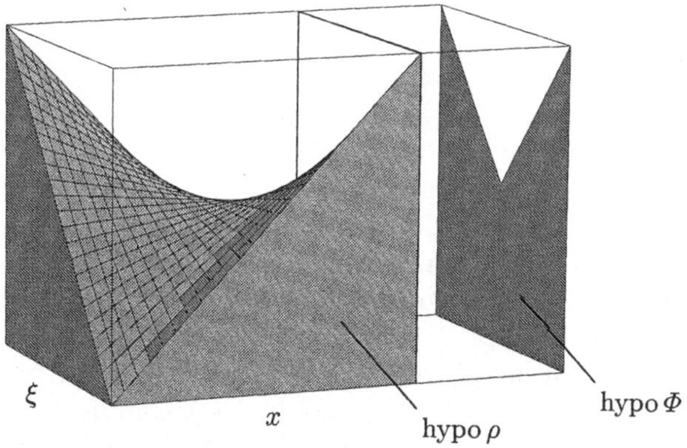

Fig. 3.2. Example of a non-concave function $\rho(x, \xi)$, which is, however, concave in each of its arguments. The corresponding sup-projection $\Phi(\xi)$ is convex, and its hypograph is given by the projection of hypo ρ along the x-axis

3.5 Saddle Structure

The following technical proposition is based on the results in Sect. 3.3 and will be needed to establish the saddle property of the recourse functions on a neighborhood of their natural domains.

Proposition 3.13. *Consider a stochastic program subject to the regularity conditions (B1)–(B5). Then, there are open convex sets*

$$\left.\begin{array}{lllr} Y^t_\cap & \subset & \mathbb{R}^{n^t} \times \mathbb{R}^{L^t} & t \in \tau \\ Y^t_\cup & \subset & \mathbb{R}^{K^t} & t \in \tau \\ Z^t_\cap & \subset & \mathbb{R}^{n^{t-1}} \times \mathbb{R}^{L^t} & t \in \tau_{-0} \\ Z^t_\cup & \subset & \mathbb{R}^{K^t} & t \in \tau_{-0} \\ Z^0_\cap & \subset & \mathbb{R}^{L_0} & \\ Z^0_\cup & \subset & \mathbb{R}^{K_0} & \end{array}\right\} \tag{3.6}$$

with the following properties:

(a) $Y^t_\cap \times Y^t_\cup$ *is a neighborhood of* Y^t, $t \in \tau$*;*

(b) $Z^t_\cap \times Z^t_\cup$ *is a neighborhood of* Z^t, $t \in \tau$*;*

(c) ρ_t *is a continuous saddle function on* $Y^t_\cap \times Y^t_\cup$ *being concave in* $\boldsymbol{x}^t$*, convex in* $\boldsymbol{\eta}^t$*, and constant in* $\boldsymbol{\xi}^t$, $t \in \tau$*;*

(d) $\boldsymbol{f}_t$ *is a continuous convex function on* $Y^t_\cap \times Y^t_\cup$ *being jointly convex in* $(\boldsymbol{x}^t, \boldsymbol{\xi}^t)$ *and constant in* $\boldsymbol{\eta}^t$, $t \in \tau$*;*

(e) $Y^t_\cap \times Y^t_\cup \times \Theta_{t+1} \times \Xi_{t+1} \subset Z^{t+1}_\cap \times Z^{t+1}_\cup$ *for all* $t \in \tau_{-T}$*;*

(f) the graph of X_t over $Z^t_\cap \times Z^t_\cup$ is a subset of $Y^t_\cap \times Y^t_\cup$, $t \in \tau$;

(g) the multifunction X_t is non-empty-compact-valued on $Z^t_\cap \times Z^t_\cup$, $t \in \tau$.

Proof. The claim is proved by backward induction. By the regularity conditions (B2) and (B3) there are open convex sets $Y^T_\cap$ and $Y^T_\cup$ which satisfy the requirements (a), (c), and (d). This observation completes the basis step. Assume now that the existence of suitable sets $Y^t_\cap$ and $Y^t_\cup$ has been shown for some $t \in \tau$. Then, the propositions 3.7, 3.9, and 3.10 ensure the existence of open convex sets $Z^t_\cap$ and $Z^t_\cup$ which satisfy the requirements (b), (f), and (g). If $t = 0$, we are finished. Otherwise, by compactness of $Z^t = Y^{t-1} \times \Theta_t \times \Xi_t$ (cf. the propositions 2.3 and 3.8) and by the regularity conditions (B2) and (B3) there are suitable sets $Y^{t-1}_\cap$ and $Y^{t-1}_\cup$ which satisfy the requirements (a), (c), (d), and (e). Therefore, the induction step is established, and the claim follows. □

Below we will assume that a collection of sets as in (3.6) is given, and that these sets exhibit the properties postulated in proposition 3.13. Now we are ready to state the main result of this section.

Theorem 3.14 (Saddle Property I). *Under the conditions (B1)–(B5) the recourse function Φ_t has a saddle structure on a convex neighborhood of Z^t, $t \in \tau$. Concretely speaking, Φ_t is concave in $(\boldsymbol{x}^{t-1}, \boldsymbol{\xi}^t)$ and convex in $\boldsymbol{\eta}^t$, $t \in \tau_{-0}$, while Φ_0 is concave in $\boldsymbol{\xi}_0$ and convex in $\boldsymbol{\eta}_0$. Moreover, the expectation functional $(E_t\Phi_{t+1})$ has a saddle structure on a convex neighborhood of Y^t, being concave in $(\boldsymbol{x}^t, \boldsymbol{\xi}^t)$ and convex in $\boldsymbol{\eta}^t$, $t \in \tau_{-T}$.*

Proof. We will argue that Φ_t has a saddle structure on $Z^t_\cap \times Z^t_\cup$ for all $t \in \tau$, whereas $(E_t\Phi_{t+1})$ has a saddle structure on $Y^t_\cap \times Y^t_\cup$ for all $t \in \tau_{-T}$. The proof is by backward induction on t. First, introduce an auxiliary objective function

$$\bar{\rho}_T := \begin{cases} \hat{\rho}_T \text{ on } & Y^T_\cap \times Y^T_\cup, \\ +\infty \text{ on } & Y^T_\cap \times (Y^T_\cup)^c, \\ -\infty \text{ everywhere else,} & \end{cases}$$

and a sup-projection

$$\bar{\Phi}_T(\boldsymbol{x}^{T-1}, \boldsymbol{\eta}^T, \boldsymbol{\xi}^T) := \sup_{\boldsymbol{x}_T \in \mathbb{R}^{n_T}} \bar{\rho}_T(\boldsymbol{x}^T, \boldsymbol{\eta}^T, \boldsymbol{\xi}^T).$$

By assumption (B2) the extended-real-valued auxiliary function $\bar{\rho}_T$ exhibits a saddle structure on $\mathbb{R}^{n^T} \times \mathbb{R}^{K^T} \times \mathbb{R}^{L^T}$, being concave in $(\boldsymbol{x}^T, \boldsymbol{\xi}^T)$ and convex in $\boldsymbol{\eta}^T$. Then, by proposition 3.11 the value function $\bar{\Phi}_T$ is convex in $\boldsymbol{\eta}^T$ for all fixed values of $(\boldsymbol{x}^{T-1}, \boldsymbol{\xi}^T)$. Moreover, proposition 3.12 implies concavity of $\bar{\Phi}_T$ in $(\boldsymbol{x}^{T-1}, \boldsymbol{\xi}^T)$ for all fixed values of $\boldsymbol{\eta}^T$. By the construction of the auxiliary objective function and by statement (f) of proposition 3.13 we have $\Phi_T \equiv \bar{\Phi}_T$

on $Z_{\cap}^T \times Z_{\cup}^T$. This implies that Φ_T shows the postulated saddle structure on the open product set $Z_{\cap}^T \times Z_{\cup}^T$. Consequently, the basis step is established.

Assume now that the recourse function of stage $t+1$ exhibits the postulated saddle property, i.e. assume that Φ_{t+1} is a saddle function on the product set $Z_{\cap}^{t+1} \times Z_{\cup}^{t+1}$. Then, by assertion (e) of proposition 3.13 we may conclude that Φ_{t+1} is a saddle function on $Y_{\cap}^t \times Y_{\cup}^t \times \Theta_{t+1} \times \Xi_{t+1}$. Next, consider the expectation functional

$$
\begin{aligned}
&(E_t\Phi_{t+1})\,(\boldsymbol{x}^t, \boldsymbol{\eta}^t, \boldsymbol{\xi}^t) \\
&= \int_{\Theta_{t+1}\times\Xi_{t+1}} \Phi_{t+1}(\boldsymbol{x}^t, \boldsymbol{\eta}^t, \boldsymbol{\eta}_{t+1}, \boldsymbol{\xi}^t, \boldsymbol{\xi}_{t+1})\, dP_{t+1}(\boldsymbol{\eta}_{t+1}, \boldsymbol{\xi}_{t+1} | \boldsymbol{\eta}^t, \boldsymbol{\xi}^t) \\
&= \int_{\mathcal{E}_{t+1}^{\mathrm{o}}\times\mathcal{E}_{t+1}^{\mathrm{r}}} \big(\Phi_{t+1}\circ H_{t+1}\big)(\boldsymbol{x}^t, \boldsymbol{\eta}^t, \boldsymbol{\varepsilon}_{t+1}^{\mathrm{o}}, \boldsymbol{\xi}^t, \boldsymbol{\varepsilon}_{t+1}^{\mathrm{r}})\, dQ_{t+1}(\boldsymbol{\varepsilon}_{t+1}^{\mathrm{o}}, \boldsymbol{\varepsilon}_{t+1}^{\mathrm{r}}).
\end{aligned}
$$

The transformation H_{t+1} is linear, and its matrix is given by (cf. (B4))

$$
H_{t+1} := \begin{bmatrix} 1_{n^t} & & & & \\ & 1_{K^t} & & & \\ & H_{t+1}^{\mathrm{o}} & 1_{K_{t+1}} & & \\ & & & 1_{L^t} & \\ & & & H_{t+1}^{\mathrm{r}} & 1_{L_{t+1}} \end{bmatrix}.
$$

Thereby, 1_{n^t} denotes the identity matrix on $\mathbb{R}^{n^t}$ etc. Notice that H_{t+1} preserves volumes and does not mix the arguments in which Φ_{t+1} is concave with those in which it is convex. This implies that $\Phi_{t+1} \circ H_{t+1}$ is a saddle function on $Y_{\cap}^t \times Y_{\cup}^t \times \mathcal{E}_{t+1}^{\mathrm{o}} \times \mathcal{E}_{t+1}^{\mathrm{r}}$. Finally, the expectation functional inherits the saddle structure of the integrand, since the operation $\int dQ_{t+1}$ can be viewed as taking a nonnegative linear combination of saddle functions over $Y_{\cap}^t \times Y_{\cup}^t$. An equivalent argument for purely convex functions can be found in Wets [108, proposition 2.1].

The assertions (c) and (d) of proposition 3.13 guarantee that the effective profit function $\hat{\rho}_t$ is jointly concave in $(\boldsymbol{x}^t, \boldsymbol{\xi}^t)$ and convex in $\boldsymbol{\eta}^t$ on $Y_{\cap}^t \times Y_{\cup}^t$. Next, introduce an auxiliary objective function (see Fig. 3.3)

$$
\bar{\rho}_t := \begin{cases} \hat{\rho}_t + (E_t\Phi_{t+1}) & \text{on} \quad Y_{\cap}^t \times Y_{\cup}^t, \\ +\infty & \text{on} \quad Y_{\cap}^t \times (Y_{\cup}^t)^{\mathrm{c}}, \\ -\infty & \text{everywhere else,} \end{cases}
$$

and a sup-projection

$$
\bar{\Phi}_t(\boldsymbol{x}^{t-1}, \boldsymbol{\eta}^t, \boldsymbol{\xi}^t) := \sup_{\boldsymbol{x}_t \in \mathbb{R}^{n_t}} \bar{\rho}_t(\boldsymbol{x}^t, \boldsymbol{\eta}^t, \boldsymbol{\xi}^t).
$$

According to our previous results both $(E_t\Phi_{t+1})$ and $\hat{\rho}_t$ are saddle functions on the convex product set $Y_{\cap}^t \times Y_{\cup}^t$. Thus, the extended-real-valued auxiliary

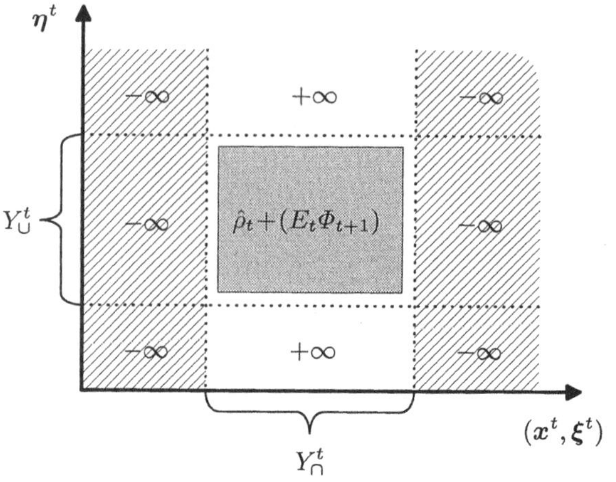

Fig. 3.3. $Y^t_\cap \times Y^t_\cup$ is a convex neighborhood of Y^t (*grey shaded area*), where both the effective profit function $\hat{\rho}_t$ and the expectation functional $(E_t\Phi_{t+1})$ have a characteristic saddle structure. Thus, the auxiliary objective function $\bar{\rho}_t$ exhibits a saddle structure on the entire underlying space

function $\bar{\rho}_t$ exhibits a saddle structure on $\mathbb{R}^{n^t} \times \mathbb{R}^{K^t} \times \mathbb{R}^{L^t}$, being concave in $(\boldsymbol{x}^t, \boldsymbol{\xi}^t)$ and convex in $\boldsymbol{\eta}^t$. Now we are prepared to apply the specific results about sup-projections. By proposition 3.11, the value function $\bar{\Phi}_t$ is convex in $\boldsymbol{\eta}^t$ for all fixed values of $(\boldsymbol{x}^{t-1}, \boldsymbol{\xi}^t)$. Moreover, proposition 3.12 implies concavity of $\bar{\Phi}_t$ in $(\boldsymbol{x}^{t-1}, \boldsymbol{\xi}^t)$ for all fixed values of $\boldsymbol{\eta}^t$. By the construction of the auxiliary objective function and by statement (f) of proposition 3.13 it is clear that $\Phi_t \equiv \bar{\Phi}_t$ on $Z^t_\cap \times Z^t_\cup$. This implies that Φ_t shows the postulated saddle structure on $Z^t_\cap \times Z^t_\cup$. Hence, the induction step is complete, and the claim follows. □

Theorem 3.14 is a slight generalization of the results in [43, Sect. 2] since convexity of the profit functions in $\boldsymbol{\eta}$ as well as concavity of the constraint functions in $\boldsymbol{\xi}$ are only required to hold on a neighborhood of Y^t instead of the entire underlying space. Moreover, in [43] the level set $\text{lev}_{\leq \mathbf{0}} \boldsymbol{f}_t$ is required to be compact. Here, however, we basically postulate compactness of the generalized feasible set Y^t, which is generically a real subset of $\text{lev}_{\leq \mathbf{0}} \boldsymbol{f}_t$ (see also Fig. 2.3). Notice that condition (B3) allows for a special class of separable constraint functions $\boldsymbol{f}_t = \boldsymbol{f}^1_t + \boldsymbol{f}^2_t$. Thereby, $\boldsymbol{f}^1_t$ is convex in $\boldsymbol{x}^t$ and constant in $\boldsymbol{\xi}^t$, whereas $\boldsymbol{f}^2_t$ is convex in $\boldsymbol{\xi}^t$ and constant in $\boldsymbol{x}^t$.

In Chap. 5 we will further investigate the dynamic versions of specific stochastic programs. Concretely speaking, we will study the Lagrangian duals of some optimization problems of the form 3.2. As in classical literature on convex analysis [89, 90], the objective functions of these maximization problems must be concave in the decision variables. This requirement is guaranteed by the following corollary 3.15, which states that the recourse functions and the expectation functionals are concave in the decision variables not only on their natural domains but on the entire underlying spaces.

Corollary 3.15. *Under the conditions (B1)–(B5), the recourse function* Φ_t *is concave in* $\boldsymbol{x}^{t-1}$ *on the entire space,* $t \in \tau_{-0}$*. Moreover, the expectation functional* $(E_t \Phi_{t+1})$ *is concave in* $\boldsymbol{x}^t$ *on the entire space,* $t \in \tau_{-T}$.

Proof. By the assumptions (B2) and (B3), the profit functions are concave whereas the constraint functions are convex in the decision variables on the underlying Euclidean spaces. Using this observation, the claim follows from a similar argument as in theorem 3.14. □

It should be remarked that condition (B2) can be relaxed to allow for $\boldsymbol{\xi}$-dependent profit functions:

(B1)' = (B1);
(B2)' – the profit function ρ_t is continuous on the underlying Euclidean space and concave in $\boldsymbol{x}^t$ for fixed values of the stochastic parameters, $t \in \tau$;
– there is a convex neighborhood of Y^t on which ρ_t is a saddle function being jointly concave in $(\boldsymbol{x}^t, \boldsymbol{\xi}^t)$ and convex in $\boldsymbol{\eta}^t$, $t \in \tau$;
(B3)' = (B3); (B4)' = (B4); (B5)' = (B5).

Corollary 3.16 (Saddle Property II). *Assume the stochastic program (3.2) to satisfy the regularity conditions (B1)'–(B5)'. Then, the conclusions of theorem 3.14 are still true.*

Proof. With obvious modifications, the proof of theorem 3.14 still applies. □

The generalized regularity conditions (B1)'–(B5)' are of minor importance for practical applications. However, it is possible to formulate a third set of regularity conditions, which allows for a different class of and constraint functions.

(B1)" = (B1);

(B2)" – the profit function ρ_t is continuous on the underlying Euclidean space and concave in $\boldsymbol{x}^t$ for fixed values of the stochastic parameters, $t \in \mathcal{T}$;
– there is a convex neighborhood of Y^t where ρ_t is a saddle function being jointly concave in $(\boldsymbol{x}_t, \boldsymbol{\xi}_t)$, concave in $\boldsymbol{x}^{t-1}$, convex in $\boldsymbol{\eta}^t$, and constant in $\boldsymbol{\xi}^{t-1}$, $t \in \mathcal{T}$;

(B3)" – the constraint function $\boldsymbol{f}_t$ is continuous on the underlying Euclidean space and convex in $\boldsymbol{x}^t$ for fixed values of the stochastic parameters, $t \in \mathcal{T}$;
– there is a convex neighborhood of Y^t where $\boldsymbol{f}_t$ is jointly convex in $(\boldsymbol{x}_t, \boldsymbol{\xi}_t)$, convex in $\boldsymbol{x}^{t-1}$ and constant in $(\boldsymbol{\eta}^t, \boldsymbol{\xi}^{t-1})$, $t \in \mathcal{T}$;
– the feasible set mapping X_t is bounded on a neighborhood of Z^t, $t \in \mathcal{T}$;

(B4)" the random data $\{\boldsymbol{\eta}_t, \boldsymbol{\xi}_t\}_{t\in\mathcal{T}}$ follows a block-diagonal autoregressive process; $\mathcal{E}_t^{\text{o}}$ and $\mathcal{E}_t^{\text{r}}$ are compact simplices, while the matrix of AR coefficients H_t^{r} equals 0 for all $t \in \mathcal{T}_{-0}$;

(B5)" = (B5).

Corollary 3.17 (Saddle Property III). *Assume the stochastic optimization problem (3.2) to fulfil the conditions (B1)"–(B5)". Then, the recourse function* Φ_t *is convex in* $\boldsymbol{\eta}^t$*, biconcave (separately concave) in* $\boldsymbol{x}^{t-1}$ *and* $\boldsymbol{\xi}_t$*, and independent of* $\boldsymbol{\xi}^{t-1}$ *on a convex neighborhood of* Z^t*,* $t \in \mathcal{T}_{-0}$*. Moreover,* Φ_0 *is convex in* $\boldsymbol{\eta}_0$ *and concave in* $\boldsymbol{\xi}_0$ *on a convex neighborhood of* Z^0*. The expectation functional* $(E_t\Phi_{t+1})$ *is concave in* $\boldsymbol{x}^t$*, convex in* $\boldsymbol{\eta}^t$*, and constant in* $\boldsymbol{\xi}^t$ *on a convex neighborhood of* Y^t*,* $t \in \mathcal{T}_{-T}$*.*

Proof. The proof of corollary 3.17 is widely parallel to that of theorem 3.14. Here, however, we exploit the fact that the expectation functionals are independent of the random parameters $\boldsymbol{\xi}$ due to the assumptions (B2)", (B3)", and (B4)".[3] □

Notice that condition (B3)" allows for a special class of separable constraint functions $\boldsymbol{f}_t = \boldsymbol{f}_t^1 + \boldsymbol{f}_t^2$. Thereby, $\boldsymbol{f}_t^1$ is constant in $(\boldsymbol{x}^{t-1}, \boldsymbol{\xi}^t)$ and convex in $\boldsymbol{x}_t$, whereas $\boldsymbol{f}_t^2$ is constant in $\boldsymbol{x}_t$ and *biconvex* (separately convex)

[3] Corollary 3.17 is closely related to theorem 1 in Birge and Louveaux [10, Sect. 11.1], which basically exploits the convexity properties of a linear stochastic program to construct a dual feasible solution.

in $\boldsymbol{x}^{t-1}$ and $\boldsymbol{\xi}^t$. Thus, corollary 3.17 allows for stochastic technology matrices in linear multistage stochastic programs.[4] On the other hand, the uncertain parameters $\boldsymbol{\xi}_t$ necessarily describe a sequence of independent random variables. They do not have the flexibility to fit empirical data with serial correlations. Because of this major drawback we will concentrate on the regularity conditions (B1)–(B5) in the remainder.

3.6 Subdifferentiability

In this section we will argue that the regularity conditions (B1)–(B5) imply subdifferentiability of the recourse functions and the expectation functionals on their natural domains. Subdifferentiability and continuity are complementary properties of convex and concave functions in the sense that none is implied by the other. For example, consider the concave function defined through

$$\Phi : \begin{cases} [0,1] \to \mathbb{R} \\ x \mapsto \sqrt{x}\,. \end{cases}$$

This Φ is continuous relative to its domain, but it is not subdifferentiable at the origin, $\partial\Phi(0) = \emptyset$. On the other hand, an example of an lsc convex function which is not continuous relative to the domain of its subdifferential mapping can be found in [89, p. 83]. However, Rockafellar shows that continuity, subdifferentiability, and pointwise finiteness are equivalent properties of a saddle function defined on an open subset of a finite-dimensional Euclidean space (cf. [89, theorem 10.1 and theorem 23.4]).

Theorem 3.18 (Subdifferentiability). *Consider a stochastic program subject to the conditions (B1)–(B5). Then, the recourse function Φ_t is subdifferentiable on a neighborhood of Z^t, $t \in \tau$. Moreover, the expectation functional $(E_t\Phi_{t+1})$ is subdifferentiable on a neighborhood of Y^t, $t \in \tau_{-T}$.*

Proof. We will argue that Φ_t is subdifferentiable and continuous on $Z^t_\cap \times Z^t_\cup$ for all indices $t \in \tau$, whereas $(E_t\Phi_{t+1})$ is subdifferentiable and continuous on $Y^t_\cap \times Y^t_\cup$ for $t \in \tau_{-T}$. As usual, the claim is proved by backward induction. First, we show that Φ_T is pointwise finite on $Z^T_\cap \times Z^T_\cup$. To this end, select a reference point $(\boldsymbol{x}^{T-1}, \boldsymbol{\eta}^T, \boldsymbol{\xi}^T) \in Z^T_\cap \times Z^T_\cup$. By assertion (g) of proposition 3.13 the corresponding stage T feasible set $X_T(\boldsymbol{x}^{T-1}, \boldsymbol{\xi}^T)$ is compact and non-empty, while the associated objective function

$$\boldsymbol{x}_T \mapsto \rho_T(\boldsymbol{x}^T, \boldsymbol{\eta}^T, \boldsymbol{\xi}^T)$$

[4] Cf. Sect. 5.4.

is continuous on $X_T(\boldsymbol{x}^{T-1}, \boldsymbol{\xi}^T)$. This follows immediately from assertion (f) of proposition 3.13. Then, the Weierstrass maximum theorem implies finiteness of $\Phi_T(\boldsymbol{x}^{T-1}, \boldsymbol{\eta}^T, \boldsymbol{\xi}^T)$. As the choice of the reference point was arbitrary, Φ_T is pointwise finite on $Z_\cap^T \times Z_\cup^T$. In the proof of theorem 3.14 we showed that Φ_T is a saddle function on the open set $Z_\cap^T \times Z_\cup^T$. Thus, pointwise finiteness ensures via [89, theorem 23.4 and theorem 10.1] that Φ_T is subdifferentiable and continuous on $Z_\cap^T \times Z_\cup^T$. Thus, the basis step is proved.

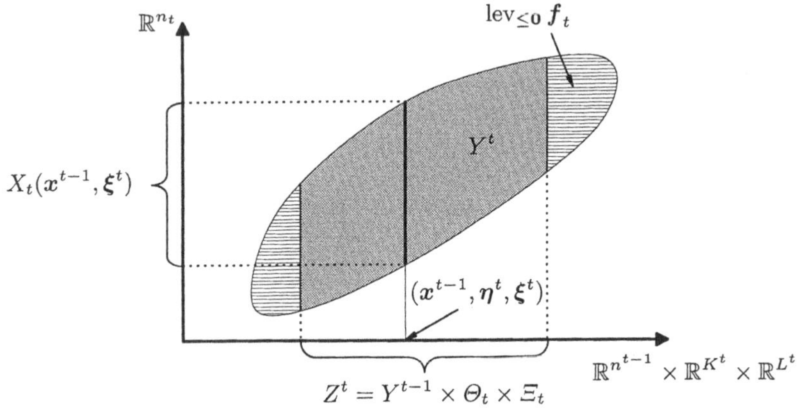

Fig. 3.4. Visualization of the generalized feasible sets Y^t and Z^t. By definition, $\text{lev}_{\leq 0}\, \boldsymbol{f}_t$ is a subset of $\mathbb{R}^{n^t} \times \mathbb{R}^{K^t} \times \mathbb{R}^{L^t}$ and a superset of Y^t. The projection of Y^t on $\mathbb{R}^{n^{t-1}} \times \mathbb{R}^{K^t} \times \mathbb{R}^{L^t}$ coincides with Z^t. Each reference point $(\boldsymbol{x}^{t-1}, \boldsymbol{\eta}^t, \boldsymbol{\xi}^t)$ in Z^t defines a fibre in Y^t (*bold line*), whose projection on $\mathbb{R}^{n_t}$ corresponds to the feasible set $X_t(\boldsymbol{x}^{t-1}, \boldsymbol{\xi}^t)$

Let us now assume that we have already shown subdifferentiability and continuity of Φ_{t+1} on $Z_\cap^{t+1} \times Z_\cup^{t+1}$. Then, it can easily be verified that the recourse function Φ_{t+1} is also continuous on $Y_\cap^t \times Y_\cup^t \times \Theta_{t+1} \times \Xi_{t+1}$, as implied by assertion (e) of proposition 3.13. Moreover, the expectation functional $(E_t\Phi_{t+1})$ is continuous and, a fortiori, pointwise finite on $Y_\cap^t \times Y_\cup^t$ by the dominated convergence theorem, continuity of the integrand, and compactness of the integration region. Likewise, assumption (B2) requires continuity of the profit function ρ_t on the entire underlying space and, a fortiori, on $Y_\cap^t \times Y_\cup^t$.

Next, select a reference point $(\boldsymbol{x}^{t-1}, \boldsymbol{\eta}^t, \boldsymbol{\xi}^t) \in Z_\cap^t \times Z_\cup^t$. By assertion (g) of proposition 3.13 the corresponding stage t feasible set $X_t(\boldsymbol{x}^{t-1}, \boldsymbol{\xi}^t)$ is compact and non-empty, while the associated objective function

$$\boldsymbol{x}_t \mapsto \rho_t(\boldsymbol{x}^t, \boldsymbol{\eta}^t, \boldsymbol{\xi}^t) + (E_t\Phi_{t+1})(\boldsymbol{x}^t, \boldsymbol{\eta}^t, \boldsymbol{\xi}^t)$$

is continuous on $X_t(\boldsymbol{x}^{t-1}, \boldsymbol{\xi}^t)$ (see also Fig. 3.4). This follows immediately from the continuity results of the previous paragraph and assertion (f) of proposition 3.13. Then, the Weierstrass maximum theorem implies finiteness of $\Phi_t(\boldsymbol{x}^{t-1}, \boldsymbol{\eta}^t, \boldsymbol{\xi}^t)$. As the choice of the reference point was arbitrary, Φ_t is pointwise finite on $Z^t_\cap \times Z^t_\cup$.

In the proof of theorem 3.14 we showed that Φ_t is a saddle function on the open set $Z^t_\cap \times Z^t_\cup$, while the expectation functional $(E_t\Phi_{t+1})$ is a saddle function on the open set $Y^t_\cap \times Y^t_\cap$. Thus, pointwise finiteness ensures via [89, theorem 23.4 and theorem 10.1] that Φ_t is subdifferentiable and continuous on $Z^t_\cap \times Z^t_\cup$ whereas $(E_t\Phi_{t+1})$ is subdifferentiable and continuous on $Y^t_\cap \times Y^t_\cup$. □

Theorem 3.18 is closely related to the stability criterion in [42, theorem 10.2]. Moreover, it provides a slight generalization of the results in [43, Sect. 2], as our definition of Slater points allows for stochastic programs with equality constraints.

4

Barycentric Approximation Scheme

4.1 Scenario Generation

The solution of stochastic programs poses severe difficulties, especially in the multistage case. If the underlying probability space is continuous, the static version of some stochastic program represents an optimization problem over an infinite-dimensional function space (see e.g. (2.8)). Then, analytical solutions are available only for unrealistically simple models. Analytical treatment of the dynamic version of a stochastic program is no less challenging. The reason for that is twofold. First, instead of one 'large' mathematical program one faces for each decision stage a parametric family of 'smaller' mathematical programs. Moreover, evaluation of the expectation functionals requires multivariate integration of a function which is only known implicitly as the result of a subordinate parametric optimization problem.

Numerical solutions are usually based on a suitable discretization of the continuous probability space. The standard approach is to select a discrete probability measure with finite support and solve the stochastic program with respect to this discrete auxiliary measure instead of the continuous original measure. In doing so, one effectively approximates the original stochastic program by an optimization problem over a finite-dimensional Euclidean space, which is numerically tractable.

The selection of an appropriate discrete probability measure is referred to as *scenario generation* and represents a primary challenge in the field of stochastic programming. For notational convenience, here, we work with the space $(\Omega, \mathcal{B}(\Omega), P)$ instead of the augmented probability space[1] introduced in Sect. 3.1. Scenario generation is based upon the following procedure, which is motivated by the exposition in [44, Sect. 3]. For each $t \in \tau_{-0}$ let P_t be a regular conditional probability for P on $\mathcal{B}(\Omega_t) \times \Omega^{t-1}$, and let P_0 be the

[1]Without loss of generality, we may assume that $\Omega = \Theta \times \Xi$.

marginal probability measure of P on $\mathcal{B}(\Omega_0)$. Assume that for each outcome history $\boldsymbol{\omega}^{t-1} \in \Omega^{t-1}$ a discrete measure $P_t^d(\cdot|\boldsymbol{\omega}^{t-1})$ approximates $P_t(\cdot|\boldsymbol{\omega}^{t-1})$ on $\mathcal{B}(\Omega_t)$, and a discrete measure P_0^d approximates P_0 on $\mathcal{B}(\Omega_0)$. Furthermore, for every $t \in \tau_{-0}$ the assignment

$$P_t^d : \begin{cases} \mathcal{B}(\Omega_t) \times \Omega^{t-1} \to [0,1] \\ (A, \boldsymbol{\omega}^{t-1}) \mapsto P_t^d(A|\boldsymbol{\omega}^{t-1}) \end{cases}$$

characterizes a *transition probability*, i.e. it complies with the following conditions:

(i) $P_t^d(\cdot|\boldsymbol{\omega}^{t-1})$ is a regular probability measure on $\mathcal{B}(\Omega_t)$ for any fixed outcome history $\boldsymbol{\omega}^{t-1} \in \Omega^{t-1}$;

(ii) $P_t^d(A|\cdot)$ is a Borel measurable function on Ω^{t-1} for every fixed Borel subset A of Ω_t.

Then, by the product measure theorem [2, Sect. 2.6], the marginal measure P_0^d and the transition probabilities P_t^d of stages $t \in \tau_{-0}$ can be combined to a unique probability measure P^d on the measurable space $(\Omega, \mathcal{B}(\Omega))$. Concretely speaking, the product measure

$$P^d := P_0^d * P_1^d * \cdots * P_T^d$$

is defined through

$$P^d(A) := \int_{\Omega_0}\int_{\Omega_1} \cdots \int_{\Omega_T} 1_A(\boldsymbol{\omega})\, dP_T^d(\omega_T|\boldsymbol{\omega}^{T-1}) \cdots dP_1^d(\omega_1|\omega_0)\, dP_0^d(\omega_0),$$

where A is an arbitrary Borel subset of Ω, and 1_A is the corresponding characteristic function. The discrete measure P^d represents an approximation of the original measure P, and the transition probabilities P_t^d can be viewed as regular conditional probabilities corresponding to P^d. If the supports

$$\mathcal{A}_t(\boldsymbol{\omega}^{t-1}) := \operatorname{supp} P_t^d(\cdot|\boldsymbol{\omega}^{t-1}) \quad \text{and} \quad \mathcal{A}_0 := \operatorname{supp} P_0^d$$

are finite sets, then the support of P^d defines a finite *scenario tree* $\mathcal{A}$.

$$\mathcal{A} := \operatorname{supp} P^d = \{\boldsymbol{\omega} \in \Omega \,|\, \omega_t \in \mathcal{A}_t(\boldsymbol{\omega}^{t-1})\, \forall t \in \tau_{-0},\, \omega_0 \in \mathcal{A}_0\}$$

Any outcome history $\boldsymbol{\omega} \in \mathcal{A}$ is called a scenario, and the associated *path probability* amounts to

$$P^d(\{\boldsymbol{\omega}\}) := P_0^d(\{\omega_0\}) \prod_{t=1}^{T} P_t^d(\{\omega_t\}|\boldsymbol{\omega}^{t-1}).$$

Let $\mathcal{A}^t$ be the projection of $\mathcal{A}$ on Ω^t. Then, the outcome histories $\boldsymbol{\omega}^t \in \mathcal{A}^t$ are referred to as *stage t nodes* of the underlying scenario tree. Moreover, the cardinality of the finite set $\mathcal{A}_t(\boldsymbol{\omega}^{t-1})$ characterizes the *branching factor* of the scenario tree at node $\boldsymbol{\omega}^{t-1} \in \mathcal{A}^{t-1}$.

It is intuitively clear that an auxiliary stochastic program on a discrete probability space can be a reasonable approximation for a stochastic program on a continuous probability space if the scenarios and the associated path probabilities are chosen suitably and if we consider enough scenarios. However, there is always a tradeoff between accuracy, requiring as many scenarios as possible, and numerical efficiency, which requires the branching of the scenario tree to be sparse. Keeping the branching factor fixed at each node of the scenario tree, the dimension of the auxiliary stochastic program grows exponentially with the number of decision stages. This effect is referred to as *curse of dimensionality* in literature [7]. Consequently, limited capacity of modern computers forces us to concentrate on few scenarios, which must be chosen with diligence. In fact, the discrete approximate probability measure P^d should at least preserve some basic properties of the original continuous probability measure P. Moreover, it is indispensable that the solution of the approximate problem can be related in some way to the solution of the original problem, i.e. the exact solution of the auxiliary problem should provide an approximate solution of the original stochastic program. Ideally, one can find a discrete probability measure such that the optimal value and the optimizer set of the auxiliary stochastic program are, in a quantitative sense, close to the optimal value and the optimizer set of the original optimization problem, respectively.

One sophisticated scenario generation technique that handles these conflicting requirements is the *barycentric approximation scheme* by Frauendorfer [42–44, 46]. This specific method exploits the structural properties of convex multistage stochastic programs on an augmented probability space, which fulfill the regularity conditions (B1)–(B5). Furthermore, the barycentric approximation scheme yields two discrete probability measures P^l and P^u with few scenarios. The optimal values of the associated auxiliary stochastic programs provide upper and lower bounds on the optimal value of the original stochastic program. Therefore, P^l and P^u are referred to as *bounding measures.*

The key element in any scenario tree construction is the discretization of the conditional probability measure $P_t(\cdot|\boldsymbol{\omega}^{t-1})$. As far as the barycentric approximation scheme is concerned, the discrete approximates $P_t^l(\cdot|\boldsymbol{\omega}^{t-1})$ and $P_t^u(\cdot|\boldsymbol{\omega}^{t-1})$ represent extremal probability measures associated with specific moment problems. Since the recourse functions of regular stochastic programs are subdifferentiable saddle functions, one can show that these extremal measures are completely determined by the distributional parameters of the original probability measure P. The underlying *moment problems* are formulated as semi-infinit linear programs on specially shaped polytopes with few vertices. Thus, the corresponding extremal measures have only few discretization points and can be used to synthesize two discrete probability measures P^l and P^u defined on sparse scenario trees. Repeated application of a simple duality argument shows that P^l and P^u are indeed bounding measures.

4.2 Approximation of Expectation Functionals

For later reference we recall here some results by Frauendorfer [42, part III]. Concretely speaking, we address the problem of approximating the expectation of a saddle function from below and from above. Our argumentation applies to an important class of convex-concave saddle functions defined on a compact probability space. Although having in mind the recourse functions of a stochastic program, we initially suppress all time indices and any dependency on the outcome and decision history in order to increase readability.

Formally speaking, let Φ be a real-valued saddle function on a probability space $(\Theta \times \Xi, \mathcal{B}(\Theta \times \Xi), P)$. As usual, $E(\cdot)$ denotes expectation with respect to P. Assume that Φ is convex in $\boldsymbol{\eta} \in \Theta$, concave in $\boldsymbol{\xi} \in \Xi$, and subdifferentiable on its entire domain. Moreover, assume that $\Theta \subset \mathbb{R}^K$ and $\Xi \subset \mathbb{R}^L$ are closed regular simplices such that $\text{supp}(P)$ is a subset of the regular *cross-simplex* $\Theta \times \Xi$; the notion of a cross-simplices has been introduced in [42, p. 8]. Denote by $\{\boldsymbol{a}_\nu\}_{\nu=0}^K$ the vertices of Θ, and let $\{\boldsymbol{b}_\mu\}_{\mu=0}^L$ be the vertices of Ξ. For the time being, the simplicial covering of the support of P seems to be arbitrary. But later it will become clear that this specific choice reduces the branching of the scenario tree to be built. A rectangular covering, on the other hand, leads to an excessive number of scenarios and results in a higher computational effort.[2]

Below it proves useful to work with *barycentric coordinates*. Then, the outcomes $\boldsymbol{\eta}$ and $\boldsymbol{\xi}$ are represented as convex combinations of the vertices

$$\boldsymbol{\eta} = \sum_{\nu=0}^{K} \lambda_\nu \, \boldsymbol{a}_\nu, \qquad \sum_{\nu=0}^{K} \lambda_\nu = 1, \tag{4.1}$$

$$\boldsymbol{\xi} = \sum_{\mu=0}^{L} \tau_\mu \, \boldsymbol{b}_\mu, \qquad \sum_{\mu=0}^{L} \tau_\mu = 1, \tag{4.2}$$

and the involved coefficients are called *barycentric weights.* These weights are nonnegative if $\boldsymbol{\eta}$ and $\boldsymbol{\xi}$ lie within Θ and Ξ, respectively. Furthermore, λ_ν and τ_μ depend on the corresponding outcomes and can be combined to vectorial functions.

$$\begin{aligned} \boldsymbol{\lambda}(\boldsymbol{\eta}) &:= (\lambda_0(\boldsymbol{\eta}), \ldots, \lambda_K(\boldsymbol{\eta})) \\ \boldsymbol{\tau}(\boldsymbol{\xi}) &:= (\tau_0(\boldsymbol{\xi}), \ldots, \tau_L(\boldsymbol{\xi})) \end{aligned}$$

[2]This was e.g. pointed out by Birge and Wets [11, p. 77].

Inverting (4.1) and (4.2) yields explicit expressions for the barycentric weights

$$\boldsymbol{\lambda}(\boldsymbol{\eta}) = S^{-1}\begin{bmatrix}1\\ \boldsymbol{\eta}\end{bmatrix}, \quad \boldsymbol{\tau}(\boldsymbol{\xi}) = T^{-1}\begin{bmatrix}1\\ \boldsymbol{\xi}\end{bmatrix}, \tag{4.3}$$

where

$$S := \begin{bmatrix}1 & 1 & \cdots & 1\\ \boldsymbol{a}_0 & \boldsymbol{a}_1 & \cdots & \boldsymbol{a}_K\end{bmatrix}, \qquad T := \begin{bmatrix}1 & 1 & \cdots & 1\\ \boldsymbol{b}_0 & \boldsymbol{b}_1 & \cdots & \boldsymbol{b}_L\end{bmatrix}.$$

Notice that $S \in \mathbb{R}^{(K+1)\times(K+1)}$ and $T \in \mathbb{R}^{(L+1)\times(L+1)}$ are regular matrices since the underlying simplices are assumed to be non-degenerate. Obviously, the barycentric weights λ_ν are linear affine functions of $\boldsymbol{\eta}$, and $\lambda_{\nu_1}(\boldsymbol{a}_{\nu_2}) = \delta_{\nu_1\nu_2}$ for all $\nu_1, \nu_2 \in \{0, \ldots, K\}$. In other words, λ_ν is the unique linear affine functional on $\mathbb{R}^K$ which adopts the value 1 at $\boldsymbol{\eta} = \boldsymbol{a}_\nu$ and vanishes at all the other vertices of Θ. Similar statements hold for the functionals τ_μ. Next, we define the matrix of the *generalized barycentric weights* as the tensor product of $\boldsymbol{\lambda}$ and $\boldsymbol{\tau}$.

$$\gamma(\boldsymbol{\eta}, \boldsymbol{\xi}) := \boldsymbol{\lambda}(\boldsymbol{\eta}) \otimes \boldsymbol{\tau}(\boldsymbol{\xi}) \in \mathbb{R}^{(K+1)\times(L+1)}$$

The matrix elements of γ are given by

$$\gamma_{\nu\mu}(\boldsymbol{\eta}, \boldsymbol{\xi}) = \lambda_\nu(\boldsymbol{\eta})\, \tau_\mu(\boldsymbol{\xi}),$$

and they are bilinear functionals of $\boldsymbol{\eta}$ and $\boldsymbol{\xi}$ due to (4.3).

With these definitions we are prepared to construct bounds on the expectation of a given saddle function Φ. For didactic reasons, our argumentation is divided into three steps. First, we consider a purely convex function and exploit Jensen's inequality [57] to find a lower bound. In a second phase, we restrict ourselves to the study of a concave function and determine an upper bound by means of the Edmundson-Madansky inequality [35, 69, 70]. Finally, combining the inequalities by Jensen and Edmundson-Madansky naturally leads to bounds on the expectation of an arbitrary saddle function.

4.2.1 Jensen's Inequality

This section is devoted to the study of degenerate saddle functions $\Phi(\boldsymbol{\eta}, \boldsymbol{\xi})$ which are constant in the second argument. Thus, without loss of generality, the parameter $\boldsymbol{\xi}$ can be disregarded, and we are left with the problem of approximating a convex mapping $\Phi(\boldsymbol{\eta})$ from below by linear affine functions. The corresponding (convex) feasible set

$$\mathcal{L} := \{L \mid L(\boldsymbol{\eta}) = \langle \boldsymbol{\alpha}, \boldsymbol{\eta}\rangle - \beta \le \Phi(\boldsymbol{\eta})\ \forall \boldsymbol{\eta} \in \Theta\}$$

contains all linear affine minorants of Φ. Moreover, the marginal measure P^{o} on $(\Theta, \mathcal{B}(\Theta))$ is defined through $P^{\mathrm{o}}(A) := P(A \times \Xi)$, $A \in \mathcal{B}(\Theta)$. For simplicity, P and P^{o} are identified below, and the superscript 'o' is omitted. The best minorant L with respect to this probability measure is determined by maximizing the expected value of L over the feasible set $\mathcal{L}$.

$$\sup_{L \in \mathcal{L}} \int_{\Theta} L(\eta)\, dP(\eta) = \sup_{\alpha,\beta} \int_{\Theta} \langle \alpha, \eta \rangle - \beta\, dP(\eta) \qquad \text{s.t.} \quad \langle \alpha, \eta \rangle - \beta \le \Phi(\eta)\ \forall \eta \in \Theta \tag{4.4}$$

Obviously, the integral in the objective of (4.4) reduces to $L(E(\eta))$ and is bounded above by $\Phi(E(\eta))$. As Φ was assumed to be subdifferentiable on Θ, there exists at least one optimal supporting function $L^{\mathrm{opt}} \in \mathcal{L}$ which satisfies the strict equality

$$L^{\mathrm{opt}}(E(\eta)) = \Phi(E(\eta)).$$

As indicated in Fig. 4.1(b), the solution L^{opt} of the primal optimization problem (4.4) is not unique if the subdifferential of Φ at $E(\eta)$ contains more than one element. However, any optimal functional L^{opt} is a lower bound for Φ and reproduces the maximal objective function value $\Phi(E(\eta))$. By linear programming duality we find a corresponding dual program (cf. [64, proposition 4.8]; for a general introduction to duality theory we refer to [90] and the summary in appendix A).

$$\inf_{Q} \int_{\Theta} \Phi(\eta) dQ(\eta) \qquad \text{s.t.} \quad Q \text{ is a probability measure on } \Theta \text{ with } \int_{\Theta} \eta\, dQ(\eta) = E(\eta) \tag{4.5}$$

The moment problem (4.5) is in fact a minimization problem over all probability measures Q conserving the expectation value (first moment) of the true measure P. For additional information about the use of generalized moment problems in stochastic programming we refer to Dupačová [26, 28]. Since Φ is convex and subdifferentiable on Θ, the dual problem (4.5) is solvable, and one specific solution P^l is given by the Dirac measure concentrated at $E(\eta)$.

$$P^l = \delta_{E(\eta)} \tag{4.6}$$

The explicit construction of L^{opt} and P^l entails stability of the primal-dual pair (4.4) and (4.5), i.e. both problems are solvable and the optimal values coincide. In Fig. 4.1, the degenerate probability density function of P^l is visualized as a narrow peak located at $E(\eta)$. However, it should be noticed that the dual solution is not necessarily unique. To see this, we recall the known fact that the optimal solutions of (4.4) and (4.5) are completely determined by primal and dual feasibility as well as *complementary slackness* (cf. [64, theorem 4.2]). In our example, the complementary slackness condition reads

$$P^l(\{\eta \in \Theta \mid L^{\text{opt}}(\eta) < \Phi(\eta)\}) = 0.$$

Thus, the extremal probability measure P^l is concentrated to those points where the optimal supporting function L^{opt} touches Φ. As illustrated in Fig. 4.1(a) and (b), the optimal dual solution is uniquely given by (4.6) if Φ is strictly convex at $E(\boldsymbol{\eta})$. However, consider a convex neighborhood U of $E(\boldsymbol{\eta})$, and assume that Φ is linear affine on U (cf. Fig. 4.1(c)). Then, the unique optimal supporting function L^{opt} touches Φ on the entire neighborhood U. Any probability measure which preserves the true expectation value and whose support is covered by U minimizes (4.5). Nevertheless, the discrete measure P^l, as specified in (4.6), always constitutes a possible solution. Remarkably, P^l is independent of the underlying function Φ. Combining the above results, we end up with the well-known *Jensen inequality*.

$$\int_\Theta \Phi(\eta)dP(\eta) \geq \int_\Theta L^{\text{opt}}(\eta)dP(\eta) = \int_\Theta \Phi(\eta)dP^l(\eta) \tag{4.7}$$

Equality of the second and the third term is due to strong duality. Notice that in most textbooks a slightly different notation is used: $E(\Phi(\boldsymbol{\eta})) \geq \Phi(E(\boldsymbol{\eta}))$.

4.2.2 Edmundson-Madansky Inequality

In a next phase we focus on a degenerate saddle function which is independent of the stochastic parameter $\boldsymbol{\eta}$. Concretely speaking, we attempt to approximate a concave function $\Phi(\boldsymbol{\xi})$ from below with hyperplanes. In complete analogy to the previous section, the set of feasible supporting functions is defined as

$$\mathcal{L} := \{L \mid L(\xi) = \langle \alpha, \xi \rangle - \beta \leq \Phi(\xi)\ \forall \xi \in \Xi\}$$

and contains all linear affine minorants of Φ. As usual, the marginal measure P^{r} on $(\Xi, \mathcal{B}(\Xi))$ is given by $P^{\text{r}}(A) := P(\Theta \times A)$, $A \in \mathcal{B}(\Xi)$. However, P and P^{r} shall be identified below, and the superscript 'r' is omitted. The best minorant with respect to this probability measure is determined by means of a semi-infinite linear program.

$$\sup_{L\in\mathcal{L}} \int_\Xi L(\xi)\, dP(\xi) = \sup_{\alpha,\beta} \int_\Xi \langle \alpha, \xi \rangle - \beta\, dP(\xi) \tag{4.8}$$
$$\text{s.t.} \quad \langle \alpha, \xi \rangle - \beta \leq \Phi(\xi)\ \forall \xi \in \Xi$$

It is intuitively clear that the optimal hyperplane L^{opt} intersects Φ at the vertices of the simplex Ξ; this implies that $\Phi(\boldsymbol{b}_\mu) = L^{\text{opt}}(\boldsymbol{b}_\mu)$ for all $\mu = 0, \ldots, L$, as sketched in Fig. 4.1(d).[3] Moreover, the primal solution is most conveniently expressed in terms of the classical barycentric weights.

[3] For a rigorous proof we refer to [41, Sect. 13].

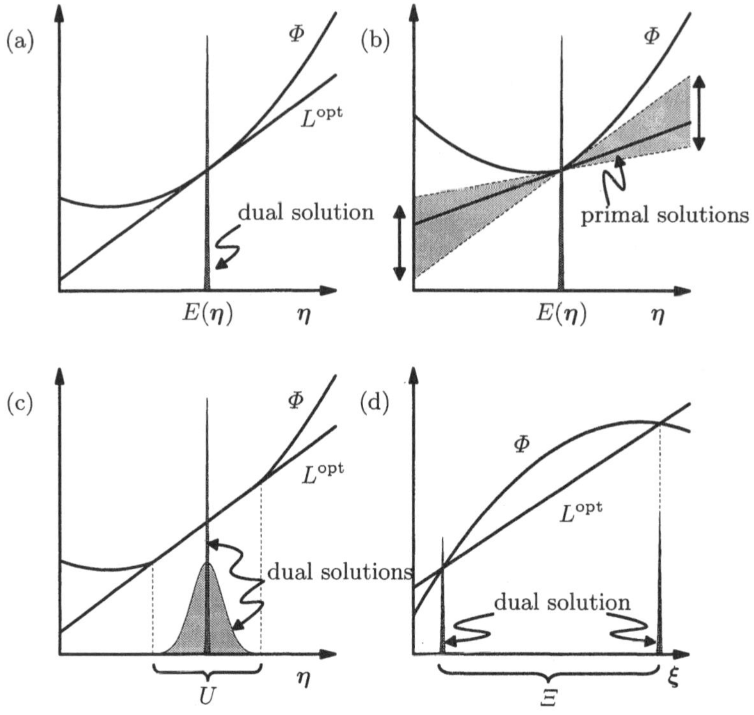

Fig. 4.1. **(a)** If $\Phi(\eta)$ is strictly convex and differentiable, the primal and dual solutions of (4.4) are unique. **(b)** In the presence of a sharp bend at $E(\eta)$, every subgradient in $\partial\Phi(E(\eta))$ corresponds to an optimal supporting function (*shaded region*). However, the extremal measure P^l remains unique. **(c)** Conversely, if $\Phi(\eta)$ is linear on a neighborhood U of $E(\eta)$, uniqueness of the dual solution is no longer assured. **(d)** As argued in Sect. 4.2.2, the extremal measure associated with a concave function $\Phi(\boldsymbol{\xi})$ is concentrated at the edges of the simplex Ξ

$$L^{\mathrm{opt}}(\boldsymbol{\xi}) = \sum_{\mu=0}^{L} \Phi(\boldsymbol{b}_\mu)\tau_\mu(\boldsymbol{\xi})$$

It is worth to be mentioned that L^{opt} is unique unless the simplex Ξ is degenerate. Little surprisingly, to the linear program (4.8) we can assign a dual moment problem.

$$\inf_Q \int_\Xi \Phi(\boldsymbol{\xi})dQ(\boldsymbol{\xi}) \tag{4.9}$$

$$\text{s.t.}\quad Q \text{ is a probability measure on } \Xi \text{ with } \int_\Xi \boldsymbol{\xi}dQ(\boldsymbol{\xi}) = E(\boldsymbol{\xi})$$

measure P^l vanishes everywhere but at the edges of Ξ, where the optimal hyperplane L^{opt} touches the concave function Φ. Thus, P^l reduces to a linear combination of Dirac measures.

$$P^l = \sum_{\mu=0}^{L} \tau_\mu(E(\boldsymbol{\xi}))\, \delta_{\boldsymbol{b}_\mu} \tag{4.10}$$

Observe that P^l is – as in the previous section – independent of Φ. Generically, (4.10) is the unique solution of the dual problem (4.9). Though, if Φ is linear affine at least on one straight line between two arbitrary vertices, then there exist additional extremal measures different from P^l. Inserting (4.10) into (4.9) yields the classical *Edmundson-Madansky inequality*.

$$\int_\Xi \Phi(\boldsymbol{\xi})dP(\boldsymbol{\xi}) \geq \int_\Xi L^{\text{opt}}(\boldsymbol{\xi})dP(\boldsymbol{\xi}) = \int_\Xi \Phi(\boldsymbol{\xi})dP^l(\boldsymbol{\xi})$$

In the next section we combine both the Jensen and Edmundson-Madansky inequalities to construct an optimal bilinear approximation for any arbitrary saddle function $\Phi(\boldsymbol{\eta}, \boldsymbol{\xi})$.

4.2.3 Lower Barycentric Approximation

Consider a subdifferentiable saddle function $\Phi(\boldsymbol{\eta}, \boldsymbol{\xi})$ on a regular cross-simplex $\Theta \times \Xi$. In analogy to the previous sections, we aim at approximating Φ from below. Since Φ is convex along Θ and concave along Ξ, it is reasonable to study specific supporting functions which are linear affine both along Θ as well as along Ξ. Thus, the space of feasible supporting functions is defined as

$$\mathcal{L} := \left\{ L \,\middle|\, L(\boldsymbol{\eta}, \boldsymbol{\xi}) = \begin{bmatrix} 1 \\ \boldsymbol{\eta} \end{bmatrix}' C \begin{bmatrix} 1 \\ \boldsymbol{\xi} \end{bmatrix} \leq \Phi(\boldsymbol{\eta}, \boldsymbol{\xi})\, \forall (\boldsymbol{\eta}, \boldsymbol{\xi}) \in \Theta \times \Xi \right\}.$$

Obviously, $\mathcal{L}$ contains all bilinear affine minorants of Φ, and each of its elements is uniquely characterized by a matrix $C \in \mathbb{R}^{(K+1)\times(L+1)}$. For the further argumentation it is convenient to give an alternative characterization of $\mathcal{L}$. Consider the set of functionals

$$\tilde{\mathcal{L}} := \Big\{ L \,\Big|\, L(\boldsymbol{\eta}, \boldsymbol{\xi}) = \sum_{\mu=0}^{L} \tau_\mu(\boldsymbol{\xi}) \left(\langle \boldsymbol{\alpha}_\mu, \boldsymbol{\eta} \rangle - \beta_\mu \right)\ \forall (\boldsymbol{\eta}, \boldsymbol{\xi}) \in \Theta \times \Xi;$$
$$\langle \boldsymbol{\alpha}_\mu, \boldsymbol{\eta} \rangle - \beta_\mu \geq \Phi(\boldsymbol{\eta}, \boldsymbol{b}_\mu)\ \forall \boldsymbol{\eta} \in \Theta\,,\ \mu = 0, \ldots, L \Big\}.$$

Below we shall argue that $\tilde{\mathcal{L}}$ coincides with the feasible set $\mathcal{L}$. By definition of the barycentric weights, any element of $\tilde{\mathcal{L}}$ can be written as

$$\sum_{\mu=0}^{L} \tau_\mu(\boldsymbol{\xi}) \left(\langle \boldsymbol{\alpha}_\mu, \boldsymbol{\eta} \rangle - \beta_\mu \right) = \begin{bmatrix} 1 \\ \boldsymbol{\eta} \end{bmatrix}' \underbrace{\begin{bmatrix} -\beta_0 & \cdots & -\beta_L \\ \boldsymbol{\alpha}_0 & \cdots & \boldsymbol{\alpha}_L \end{bmatrix}}_{=:A} T^{-1} \begin{bmatrix} 1 \\ \boldsymbol{\xi} \end{bmatrix}.$$

Consequently, each element of $\tilde{\mathcal{L}}$ is uniquely determined by some matrix $A \in \mathbb{R}^{(K+1)\times(L+1)}$ and corresponds to an element in $\mathcal{L}$ with $C = AT^{-1}$. Since T is a regular matrix, backtransformation is straightforward, $A = CT$. Furthermore, it can easily be seen that $\tilde{\mathcal{L}}$ contains only minorants of Φ.

$$\begin{aligned} L(\boldsymbol{\eta}, \boldsymbol{\xi}) &= \sum_{\mu=0}^{L} \tau_\mu(\boldsymbol{\xi}) \left(\langle \boldsymbol{\alpha}_\mu, \boldsymbol{\eta} \rangle - \beta_\mu \right) \\ &\leq \sum_{\mu=0}^{L} \tau_\mu(\boldsymbol{\xi})\, \Phi(\boldsymbol{\eta}, \boldsymbol{b}_\mu) && \text{(by definition of } \tilde{\mathcal{L}}) \\ &\leq \Phi\Big(\boldsymbol{\eta}, \sum_{\mu=0}^{L} \tau_\mu(\boldsymbol{\xi})\, \boldsymbol{b}_\mu\Big) = \Phi(\boldsymbol{\eta}, \boldsymbol{\xi}) && \text{(concavity of } \Phi \text{ in } \boldsymbol{\xi}) \end{aligned}$$

Collecting the above results, it is obvious that $\tilde{\mathcal{L}}$ contains only bilinear affine minorants of Φ, $\tilde{\mathcal{L}} \subset \mathcal{L}$. Conversely, from the definition of $\tilde{\mathcal{L}}$ it is obvious that $\mathcal{L} \subset \tilde{\mathcal{L}}$; this observation proves equality of the two sets under consideration. Let us now turn towards the determination of the best bilinear affine minorant of Φ with respect to the probability measure P. To this end, we study again a semi-infinite linear program.

$$\sup_{L \in \mathcal{L}} \int_{\Theta \times \Xi} L(\boldsymbol{\eta}, \boldsymbol{\xi})\, dP(\boldsymbol{\eta}, \boldsymbol{\xi}) \tag{4.11}$$

In order to solve (4.11), we rewrite its objective function with the help of some auxiliary probability measures $\tilde{P}_\mu$ being defined on $\Theta \times \{\boldsymbol{b}_\mu\}$, respectively.

$$\begin{aligned} \int_{\Theta \times \Xi} L(\boldsymbol{\eta}, \boldsymbol{\xi})\, dP(\boldsymbol{\eta}, \boldsymbol{\xi}) &= \int_{\Theta \times \Xi} \sum_{\mu=0}^{L} \tau_\mu(\boldsymbol{\xi}) (\langle \boldsymbol{\alpha}_\mu, \boldsymbol{\eta} \rangle - \beta_\mu)\, dP(\boldsymbol{\eta}, \boldsymbol{\xi}) \\ &= \sum_{\mu=0}^{L} q_\Xi(\mu) \int_\Theta (\langle \boldsymbol{\alpha}_\mu, \boldsymbol{\eta} \rangle - \beta_\mu) \underbrace{\int_\Xi \frac{\tau_\mu(\boldsymbol{\xi})}{q_\Xi(\mu)} dP(\boldsymbol{\eta}, \boldsymbol{\xi})}_{=:d\tilde{P}_\mu(\boldsymbol{\eta})} \end{aligned} \tag{4.12}$$

Thereby, a sequence of $L+1$ nonnegative *pseudo-probabilities* $q_\Xi(\mu)$, $\mu = 0, \ldots, L$, has been introduced.

$$q_\Xi(\mu) := \int_{\Theta \times \Xi} \tau_\mu(\boldsymbol{\xi})\, dP(\boldsymbol{\eta}, \boldsymbol{\xi}) \geq 0$$

Since the barycentric weights τ_μ sum up to unity, we find $\sum_{\mu=0}^{L} q_\Xi(\mu) = 1$. Moreover, by the definition of $q_\Xi(\mu)$ the auxiliary measures $\tilde{P}_\mu$ are normalized. Substitution of the rearranged objective function (4.12) into (4.11) yields

$$\sup_{L\in\mathcal{L}} \int_{\Theta\times\Xi} L(\boldsymbol{\eta},\boldsymbol{\xi})\, dP(\boldsymbol{\eta},\boldsymbol{\xi})$$
$$= \sum_{\mu=0}^{L} q_\Xi(\mu) \sup \int_\Theta (\langle \boldsymbol{\alpha}_\mu, \boldsymbol{\eta}\rangle - \beta_\mu) d\tilde{P}_\mu(\boldsymbol{\eta})$$
$$\text{s.t.} \quad \langle \boldsymbol{\alpha}_\mu, \boldsymbol{\eta}\rangle - \beta_\mu \le \Phi(\boldsymbol{\eta}, \boldsymbol{b}_\mu) \quad \forall \boldsymbol{\eta}\in\Theta.$$

Hence, for every vertex $\boldsymbol{b}_\mu$ of Ξ we have to find the optimal minorant of $\Phi(\cdot, \boldsymbol{b}_\mu)$ with respect to the auxiliary measure $\tilde{P}_\mu$. This problem as well as its dual counterpart have already been solved in Sect. 4.2.1. Instead of the unbiased expectation value $E(\boldsymbol{\eta})$ considered there, we are now supposed to work with the so-called *generalized barycenters* $\boldsymbol{\eta}_\mu$, $\mu = 0, \dots, L$.

$$\boldsymbol{\eta}_\mu := \int_\Theta \boldsymbol{\eta}\, d\tilde{P}_\mu(\boldsymbol{\eta}) = \sum_{\nu=0}^{K} \boldsymbol{a}_\nu \underbrace{\int_{\Theta\times\Xi} \frac{\gamma_{\nu,\mu}(\boldsymbol{\eta},\boldsymbol{\xi})}{q_\Xi(\mu)}\, dP(\boldsymbol{\eta},\boldsymbol{\xi})}_{=\lambda_\nu(\boldsymbol{\eta}_\mu)} \tag{4.13}$$

Apparently, integration of the generalized barycentric weights over the augmented probability space and normalization with $q_\Xi(\mu)$ yields the classical barycentric weights of the generalized barycenters. Remember that the generalized barycentric weights are bilinear affine in the outcomes $\boldsymbol{\eta}$ and $\boldsymbol{\xi}$, and therefore the barycenters $\boldsymbol{\eta}_\mu$ are completely determined by the cross-moments of the stochastic parameters $\boldsymbol{\eta}$ and $\boldsymbol{\xi}$. For later use we define $m \in \mathbb{R}^{(K+1)\times(L+1)}$ as the matrix of cross moments, and its entries are denoted by $m_{\nu\mu}$, $\nu = 0, \dots, K$ and $\mu = 0, \dots, L$.

$$m := \int_{\Theta\times\Xi} \begin{bmatrix} 1 \\ \boldsymbol{\eta} \end{bmatrix} \begin{bmatrix} 1 \\ \boldsymbol{\xi} \end{bmatrix}' dP(\boldsymbol{\eta},\boldsymbol{\xi})$$

Notice that $m_{00} = 1$ since P is a probability measure, which distributes unit mass on $\Theta \times \Xi$. After all, the generalized barycenters provide an explicit solution of the linear program (4.11).

$$\sup_{L\in\mathcal{L}} \int_{\Theta\times\Xi} L(\boldsymbol{\eta},\boldsymbol{\xi})\, dP(\boldsymbol{\eta},\boldsymbol{\xi}) = \sum_{\mu=0}^{L} q_\Xi(\mu)\, \Phi(\boldsymbol{\eta}_\mu, \boldsymbol{b}_\mu) \tag{4.14}$$

As outlined in the discussion of Jensen's inequality, the optimal supporting function $L^{\text{opt}}(\boldsymbol{\eta},\boldsymbol{\xi})$ is not necessarily unique since the subdifferential of $\Phi(\cdot, \boldsymbol{b}_\mu)$ may contain more than one subgradient at $\boldsymbol{\eta}_\mu$.

Like in the context of the Jensen and Edmundson-Madansky inequalities, it is worthwhile to investigate the dual counterpart of the linear program (4.11).

$$\begin{aligned} &\inf_Q \int_{\Theta\times\Xi} \Phi(\boldsymbol{\eta},\boldsymbol{\xi}) dQ(\boldsymbol{\eta},\boldsymbol{\xi}) \\ &\text{s.t.} \quad Q \text{ is a probability measure on } \Theta\times\Xi \text{ with} \\ &\int_{\Theta\times\Xi} \begin{bmatrix} 1 \\ \boldsymbol{\eta} \end{bmatrix} \begin{bmatrix} 1 \\ \boldsymbol{\xi} \end{bmatrix}' dQ(\boldsymbol{\eta},\boldsymbol{\xi}) = m \end{aligned} \tag{4.15}$$

The dual program (4.15) constitutes a minimization problem over all probability measures on $\Theta \times \Xi$ preserving the cross moments $m_{\nu\mu}$. Taking account of complementary slackness, we make a plausible ansatz for the optimal solution P^l of the moment problem (4.15).

$$P^l = \sum_{\mu=0}^{L} q_\Xi(\mu)\, \delta_{(\boldsymbol{\eta}_\mu, \boldsymbol{b}_\mu)} \,. \tag{4.16}$$

Thus, P^l has a discrete support, $\operatorname{supp} P^l = \{(\boldsymbol{\eta}_\mu, \boldsymbol{b}_\mu) | \mu = 0, \ldots, L{+}1\}$, and the probability associated with a specific atom $(\boldsymbol{\eta}_\mu, \boldsymbol{b}_\mu)$ is given by $q_\Xi(\mu)$. Much like in the previous sections, P^l is concentrated to those points where the optimal bilinear supporting function L^{opt} touches the saddle function Φ. By plugging (4.16) into (4.15) and comparing with (4.14) it can be verified that P^l is in fact a valid solution of the dual problem.[4] Notice that P^l is completely determined by the cross moments $m_{\nu\mu}$. A fortiori, it is independent of the saddle function Φ. In summary, we can derive a useful inequality that closely resembles the inequalities by Jensen and Edmundson-Madansky.

$$\int \Phi(\boldsymbol{\eta}, \boldsymbol{\xi}) dP(\boldsymbol{\eta}, \boldsymbol{\xi}) \geq \int L^{\text{opt}}(\boldsymbol{\eta}, \boldsymbol{\xi}) dP(\boldsymbol{\eta}, \boldsymbol{\xi}) = \int \Phi(\boldsymbol{\eta}, \boldsymbol{\xi}) dP^l(\boldsymbol{\eta}, \boldsymbol{\xi}) \tag{4.17}$$

The right hand side of (4.17) is called *lower barycentric approximation* of the expectation functional $(E\Phi) := \int \Phi(\boldsymbol{\eta}, \boldsymbol{\xi}) dP(\boldsymbol{\eta}, \boldsymbol{\xi})$. In fact, the discrete extremal probability measure P^l and the lower barycentric approximation are crucial ingredients of the scenario generation method for multistage stochastic programs which will be presented below.

4.2.4 Upper Barycentric Approximation

By symmetry, the problem of finding an optimal upper bound for $(E\Phi)$ can easily be reduced to the problem studied in the previous section. As the argumentation remains essentially the same, it is sufficient to present the results. For instance, the set of all bilinear affine majorants of Φ is given by

$$\begin{aligned} \mathcal{U} := \Big\{ U \,\Big|\, & U(\boldsymbol{\eta}, \boldsymbol{\xi}) = \sum_{\nu=0}^{K} \lambda_\nu(\boldsymbol{\eta}) \left(\langle \boldsymbol{\alpha}_\nu, \boldsymbol{\xi} \rangle - \beta_\nu \right) \ \forall (\boldsymbol{\eta}, \boldsymbol{\xi}) \in \Theta \times \Xi; \\ & \langle \boldsymbol{\alpha}_\nu, \boldsymbol{\xi} \rangle - \beta_\nu \geq \Phi(\boldsymbol{a}_\nu, \boldsymbol{\xi}) \ \forall \boldsymbol{\xi} \in \Xi \,,\ \nu = 0, \ldots, K \Big\} . \end{aligned}$$

The best majorant of the saddle function Φ is evaluated by means of the following semi-infinite linear program:

$$\inf_{U \in \mathcal{U}} \int_{\Theta \times \Xi} U(\boldsymbol{\eta}, \boldsymbol{\xi}) \; dP(\boldsymbol{\eta}, \boldsymbol{\xi}). \tag{4.18}$$

[4] More details are provided in [42, Sect. 14].

The corresponding dual problem reads

$$\begin{aligned}
&\sup_{Q} \int_{\Theta \times \Xi} \Phi(\eta, \xi) dQ(\eta, \xi) \\
\text{s.t.}\quad & Q \text{ is a probability measure on } \Theta \times \Xi \text{ with} \\
& \int_{\Theta \times \Xi} \begin{bmatrix} 1 \\ \eta \end{bmatrix} \begin{bmatrix} 1 \\ \xi \end{bmatrix}' dQ(\eta, \xi) = m.
\end{aligned} \tag{4.19}$$

As in the previous section, we introduce a sequence of nonnegative *pseudo-probabilities* $q_\Theta(\nu)$.

$$q_\Theta(\nu) := \int_{\Theta \times \Xi} \lambda_\nu(\eta)\; dP(\eta, \xi) \geq 0$$

In analogy to (4.13), the *generalized barycenters* are defined as

$$\xi_\nu := \sum_{\mu=0}^{L} b_\mu \underbrace{\int_{\Theta \times \Xi} \frac{\gamma_{\nu,\mu}(\eta, \xi)}{q_\Theta(\nu)}\; dP(\eta, \xi)}_{=\tau_\mu(\xi_\nu)}.$$

The barycenters as well as the pseudo-probabilities only depend on the cross moments $m_{\nu\mu}$, and they completely determine the extremal measure P^u which solves the moment problem (4.19).

$$P^u = \sum_{\nu=0}^{K} q_\Theta(\nu)\, \delta_{(a_\nu, \xi_\nu)}\,. \tag{4.20}$$

The discrete support of P^u is given by $\operatorname{supp} P^u = \{(a_\nu, \xi_\nu) | \nu = 0, \ldots, K+1\}$, and the probability associated with a specific atom (a_ν, ξ_ν) amounts to $q_\Theta(\nu)$. By strong duality, we end up with the requested inequality providing an upper bound for the expectation functional.

$$\int \Phi(\eta, \xi) dP(\eta, \xi) \leq \int U^{\text{opt}}(\eta, \xi) dP(\eta, \xi) = \int \Phi(\eta, \xi) dP^u(\eta, \xi) \tag{4.21}$$

We refer to the right hand side of (4.21) as *upper barycentric approximation* of the expectation functional $(E\Phi)$.

4.3 Partitioning

Depending on the curvature of the saddle function Φ and the range of the random vectors η and ξ, the estimates (4.17) and (4.21) for the expectation functional $(E\Phi)$ may be very coarse. If these estimates are unsatisfactory, the

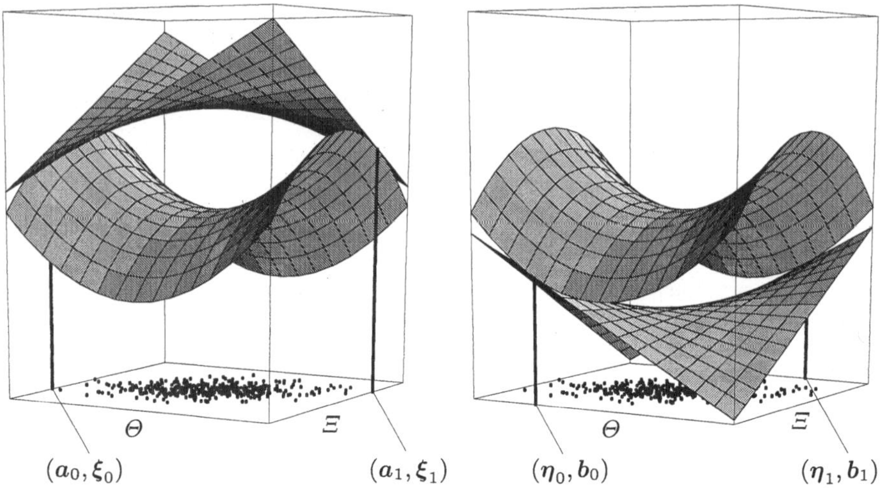

Fig. 4.2. Bilinear affine approximations of a saddle function over a two-dimensional cross-simplex $\Theta \times \Xi$

probability measure P can be represented as a convex combination of specific probability measures with a smaller support. Then, barycentric discretization is applied to each of these measures separately. Formally speaking, let $P_{J,i}$ be a regular probability measure on the augmented probability space $(\Theta \times \Xi, \mathcal{B}(\Theta \times \Xi))$ for all $i = 1, \ldots, I(J)$ such that

$$P = \sum_{i=1}^{I(J)} \varrho_{J,i} P_{J,i}, \quad \sum_{i=1}^{I(J)} \varrho_{J,i} = 1, \quad \text{and} \quad \varrho_{J,i} \geq 0 \quad \forall i = 1, \ldots, I(J). \tag{4.22}$$

The decomposition (4.22) is termed a *partition* of the measure P, and J will be referred to as the *refinement parameter*. Moreover, for every admissible combination of J and i let $\Theta_{J,i} \times \Xi_{J,i} \subset \Theta \times \Xi$ be a regular cross-simplex which covers the support of $P_{J,i}$. Then, each probability measure $P_{J,i}$ can be discretized via barycentric approximation developed in the previous sections. Concretely speaking, there are two discrete extremal probability measures $P^l_{J,i}$ and $P^u_{J,i}$ on the Borel space $(\Theta_{J,i} \times \Xi_{J,i}, \mathcal{B}(\Theta_{J,i} \times \Xi_{J,i}))$ with the same cross moments as the original measure $P_{J,i}$ such that

$$\int \Phi(\boldsymbol{\eta}, \boldsymbol{\xi}) dP^l_{J,i}(\boldsymbol{\eta}, \boldsymbol{\xi}) \leq \int \Phi(\boldsymbol{\eta}, \boldsymbol{\xi}) dP_{J,i}(\boldsymbol{\eta}, \boldsymbol{\xi}) \leq \int \Phi(\boldsymbol{\eta}, \boldsymbol{\xi}) dP^u_{J,i}(\boldsymbol{\eta}, \boldsymbol{\xi})$$

for all subdifferentiable saddle functions Φ. The discrete measures $P^l_{J,i}$ and $P^u_{J,i}$ are the solutions of the moment problems (4.15) and (4.19), where P and $\Theta \times \Xi$ are substituted by $P_{J,i}$ and $\Theta_{J,i} \times \Xi_{J,i}$, respectively. A partition of the form (4.22) always exists. For instance, suppose that $\{\hat{\Theta}_{J,i} \times \hat{\Xi}_{J,i} \,|\, i = 1, \ldots, I(J)\}$ is a disjoint *set-partition* of the augmented probability space $\Theta \times \Xi$, i.e.

$$\Theta \times \Xi = \bigcup_{i=1}^{I(J)} \hat{\Theta}_{J,i} \times \hat{\Xi}_{J,i}, \quad (\hat{\Theta}_{J,i} \times \hat{\Xi}_{J,i}) \cap (\hat{\Theta}_{J,i'} \times \hat{\Xi}_{J,i'}) = \emptyset \quad \forall i \neq i'.$$

Without much loss of generality, we may assume that $\hat{\Theta}_{J,i} \times \hat{\Xi}_{J,i}$ has non-zero probability mass, and its closure represents a compact regular cross-simplex.[5] Then, the probability measure $P_{J,i}$ can be defined through

$$P_{J,i}(A) := \frac{P(A \cap (\hat{\Theta}_{J,i} \times \hat{\Xi}_{J,i}))}{P(\hat{\Theta}_{J,i} \times \hat{\Xi}_{J,i})} \quad \forall A \in \mathcal{B}(\Theta \times \Xi).$$

Furthermore, set $\varrho_{J,i} := P(\hat{\Theta}_{J,i} \times \hat{\Xi}_{J,i})$ and $\Theta_{J,i} \times \Xi_{J,i} := \mathrm{cl}\,(\hat{\Theta}_{J,i} \times \hat{\Xi}_{J,i})$. These specifications are apparently in accordance with the general requirements (4.22). In general, for every partition of the form (4.22) one can define two discrete measures

$$P_J^l := \sum_{i=1}^{I(J)} \varrho_{J,i} P_{J,i}^l \quad \text{and} \quad P_J^u := \sum_{i=1}^{I(J)} \varrho_{J,i} P_{J,i}^u$$

on the augmented probability space $\Theta \times \Xi$. The following chain of inequalities is straightforward from the construction of the extremal probability measures $P_{J,i}^l$ and $P_{J,i}^u$.

$$\int \Phi(\eta, \xi) dP_J^l(\eta, \xi) \leq \int \Phi(\eta, \xi) dP(\eta, \xi) \leq \int \Phi(\eta, \xi) dP_J^u(\eta, \xi)$$

One may generate an 'improved' partition of the probability measure P by partitioning some or all components $P_{J,i}$ of some initial partition indexed by the given parameter J. The nested partition, which will be indexed by $J+1$, is usually referred to as a *refinement* of partition J. By successively refining the partition of P, thereby increasing the number $I(J)$ of components, one can construct two sequences of discrete probability measures $\{P_J^l\}_{J \in \mathbb{N}}$ and $\{P_J^u\}_{J \in \mathbb{N}}$ approximating the original measure P.

Proposition 4.1 (Monotonicity). *Both sequences $\{P_J^l\}_{J \in \mathbb{N}}$ and $\{P_J^u\}_{J \in \mathbb{N}}$ corresponding to a successively refined partition of the original measure P are monotonous in the following sense:*

$$\int_{\Theta \times \Xi} \Phi(\eta, \xi) dP_J^l(\eta, \xi) \geq \int_{\Theta \times \Xi} \Phi(\eta, \xi) dP_{J'}^l(\eta, \xi) \quad \forall J \geq J'$$

and

$$\int_{\Theta \times \Xi} \Phi(\eta, \xi) dP_J^u(\eta, \xi) \leq \int_{\Theta \times \Xi} \Phi(\eta, \xi) dP_{J'}^u(\eta, \xi) \quad \forall J \geq J'$$

for all subdifferentiable saddle functions Φ on $\Theta \times \Xi$.

[5] Such set-partitions are very popular in stochastic programming literature, see e.g. Huang et al. [56], Birge and Wets [11], or Kall et al. [45,60].

Proof. (Cf. Frauendorfer [42, theorem 16.2]) Without loss of generality, we investigate the sequence of lower probability measures $\{P_J^l\}_{J\in\mathbb{N}}$. Furthermore, it suffices to prove the assertion for the special case $I(J')=1$. Then, $P_{J'}^l$ is given by the extremal measure (4.16) and corresponds to a trivial partition of the original measure P. In the remainder, we shall make use of the strong duality result discussed in Sect. 4.2.3. As usual, set

$$\mathcal{L} := \{L \,|\, L \text{ is bilinear affine on } \Theta\times\Xi,\ L\le\Phi \text{ on } \Theta\times\Xi\},$$

and for all $i=1,\dots,I(J)$ define

$$\mathcal{L}_{J,i} := \{L \,|\, L \text{ is bilinear affine on } \Theta\times\Xi,\ L\le\Phi \text{ on } \Theta_{J,i}\times\Xi_{J,i}\}.$$

Then, the expectation of any subdifferentiable saddle function Φ with respect to the probability measure $P_{J'}^l$ corresponds to the optimal value of the following semi-infinite linear program

$$\int \Phi(\boldsymbol{\eta},\boldsymbol{\xi})dP_{J'}^l(\boldsymbol{\eta},\boldsymbol{\xi}) = \sup_{L\in\mathcal{L}} \int L(\boldsymbol{\eta},\boldsymbol{\xi})dP(\boldsymbol{\eta},\boldsymbol{\xi}). \tag{4.23}$$

Similarly, the expectation of Φ with respect to P_J^l amounts to

$$\int \Phi(\boldsymbol{\eta},\boldsymbol{\xi})dP_J^l(\boldsymbol{\eta},\boldsymbol{\xi}) = \sum_{i=1}^{I(J)} \varrho_{J,i} \sup_{L_{J,i}\in\mathcal{L}_{J,i}} \int L_{J,i}(\boldsymbol{\eta},\boldsymbol{\xi})dP_{J,i}(\boldsymbol{\eta},\boldsymbol{\xi}). \tag{4.24}$$

In a next step we have to show that (4.23) is smaller or equal to (4.24). To this end, assume that L is a bilinear affine functional feasible in (4.23). Then, for every index i set $L_{J,i} := L$. One can easily verify that these functionals are feasible in (4.24) and have the same objective function value as L in (4.23). Thus, expectation of Φ with respect to P_J^l is at least as large as expectation with respect to $P_{J'}^l$, yielding monotonicity of the lower expectation functionals due to refinements. An analogous argument holds for the sequence of upper measures. □

Definition 4.2. *A sequence $\{P_J\}_{J\in\mathbb{N}}$ of probability measures on the augmented probability space $(\Theta\times\Xi,\mathcal{B}(\Theta\times\Xi))$ is said to converge weakly to a probability measure P if and only if*

$$\lim_{J\to\infty}\int_{\Theta\times\Xi}\Phi(\boldsymbol{\eta},\boldsymbol{\xi})dP_J(\boldsymbol{\eta},\boldsymbol{\xi}) = \int_{\Theta\times\Xi}\Phi(\boldsymbol{\eta},\boldsymbol{\xi})dP(\boldsymbol{\eta},\boldsymbol{\xi})$$

for all continuous functions Φ on $\Theta\times\Xi$.

Proposition 4.3 (Convergence). *The sequences $\{P_J^l\}_{J\in\mathbb{N}}$ and $\{P_J^u\}_{J\in\mathbb{N}}$ converge weakly to the probability measure P provided that the diameters of all cross-simplices $\{\Theta_{J,i}\times\Xi_{J,i}\,|\,i=1,\dots,I(J)\}$ become arbitrarily small when J is increased.*

Proof. Without loss of generality we focus on the sequence of lower probability measures $\{P_J^l\}_{J\in\mathbb{N}}$. Choose any continuous function Φ on $\Theta \times \Xi$. By compactness, Φ is uniformly continuous on its domain, i.e. for every tolerance $\varepsilon > 0$ there exists a real number $\delta > 0$ such that

$$|\Phi(\eta,\xi) - \Phi(\eta',\xi')| \leq \varepsilon \quad \forall (\eta,\xi),(\eta',\xi') \in \Theta \times \Xi \\ \text{with} \quad \|(\eta,\xi) - (\eta',\xi')\| \leq \delta.$$

Next, choose J large enough to guarantee the inequality

$$\operatorname{diam}(\Theta_{J,i} \times \Xi_{J,i}) \leq \delta \quad \forall i = 1,\dots,I(J).$$

Then, we have

$$\left| \int \Phi(\eta,\xi) dP_{J,i}(\eta,\xi) - \int \Phi(\eta,\xi) dP_{J,i}^l(\eta,\xi) \right| \\ \leq \sup\{\Phi(\eta,\xi) \,|\, (\eta,\xi) \in \Theta_{J,i} \times \Xi_{J,i}\} - \inf\{\Phi(\eta,\xi) \,|\, (\eta,\xi) \in \Theta_{J,i} \times \Xi_{J,i}\}.$$

The last expression is smaller than ε due to uniform continuity of Φ. This simple estimate implies that

$$\left| \int \Phi(\eta,\xi) dP(\eta,\xi) - \int \Phi(\eta,\xi) dP_J^l(\eta,\xi) \right| \\ \leq \sum_{i=1}^{I(J)} \varrho_{J,i} \underbrace{\left| \int \Phi(\eta,\xi) dP_{J,i}(\eta,\xi) - \int \Phi(\eta,\xi) dP_{J,i}^l(\eta,\xi) \right|}_{\leq \varepsilon} \leq \varepsilon,$$

and therefore P_J^l converges weakly to the original probability measure P. □

4.4 Barycentric Scenario Trees

In this section we will construct a 'lower' ('upper') discrete probability measure P^l (P^u) on the augmented probability space of a multistage stochastic program. By replacing the original measure P with the discrete measure P^l or P^u in (3.1), an auxiliary stochastic program is generated, which allows for numerical solution. We study the structural properties of these auxiliary optimization problems under specific regularity conditions. In particular, we investigate the relationship between the recourse functions of the original and the auxiliary stochastic programs. Under certain circumstances, the auxiliary problems associated with the measures P^l and P^u provide lower and upper bounds on the optimal value of the original problem.

Consider a multistage stochastic program on an augmented probability space which satisfies at least the basic regularity conditions (B1) and (B4).

Throughout this chapter we will explicitly use the standard assumption that the random outcomes at stage 0 are deterministic. Thus, we consider η_0 and ξ_0 as fixed parameters implying that P_0 reduces to a Dirac measure. Next, for every time index $t \in \tau_{-0}$ and for any *refinement parameter* $J \in \mathbb{N}$ we establish a *partition* of the conditional probability P_t by representing it as a combination of suitable *transition probabilities*.

$$P_t(\cdot|\eta^{t-1}, \xi^{t-1}) = \sum_{i_t=1}^{I_t(J)} \varrho_{t,J,i_t}(\eta^{t-1}, \xi^{t-1}) P_{t,J,i_t}(\cdot|\eta^{t-1}, \xi^{t-1}) \tag{4.25}$$

For any admissible combination of indices t, J, and i_t, we require the transition probability $P_{t,J,i_t} : \mathcal{B}(\Theta_t \times \Xi_t) \times \Theta^{t-1} \times \Xi^{t-1} \to [0,1]$ to comply with the following conditions (see also Sect. 4.1):

(i) $P_{t,J,i_t}(\cdot|\eta^{t-1}, \xi^{t-1})$ is a regular probability measure on $\mathcal{B}(\Theta_t \times \Xi_t)$ for any fixed outcome history $(\eta^{t-1}, \xi^{t-1}) \in \Theta^{t-1} \times \Xi^{t-1}$;

(ii) $P_{t,J,i_t}(A|\cdot)$ is a Borel measurable function on $\Theta^{t-1} \times \Xi^{t-1}$ for every fixed Borel subset A of $\Theta_t \times \Xi_t$.

Furthermore, the measurable function $\varrho_{t,J,i_t} : \Theta^{t-1} \times \Xi^{t-1} \to \mathbb{R}$ is nonnegative, and we postulate

$$\sum_{i_t=1}^{I_t(J)} \varrho_{t,J,i_t} = 1 \quad \forall t \in \tau_{-0},\ J \in \mathbb{N}. \tag{4.26}$$

In a next step, for every admissible combination of t, J, and i_t we define two families of closed regular simplices in $\mathbb{R}^{K_t}$ and $\mathbb{R}^{L_t}$ as the convex hull of their vertices.

$$\begin{aligned} \Theta_{t,J,i_t}(\eta^{t-1}, \xi^{t-1}) &= \operatorname{co}\left\{ \boldsymbol{a}_{\nu_t,J,i_t}(\eta^{t-1}, \xi^{t-1}) \,\middle|\, \nu_t = 0, \ldots, K_t \right\} \\ \Xi_{t,J,i_t}(\eta^{t-1}, \xi^{t-1}) &= \operatorname{co}\left\{ \boldsymbol{b}_{\mu_t,J,i_t}(\eta^{t-1}, \xi^{t-1}) \,\middle|\, \mu_t = 0, \ldots, L_t \right\} \end{aligned} \tag{4.27}$$

Assume that the vertices $\boldsymbol{a}_{\nu_t,J,i_t}$ and $\boldsymbol{b}_{\mu_t,J,i_t}$ are measurable vector-valued functions on $\Theta^{t-1} \times \Xi^{t-1}$. The Cartesian product of the simplices (4.27) constitutes a regular cross-simplex

$$\Theta_{t,J,i_t}(\eta^{t-1}, \xi^{t-1}) \times \Xi_{t,J,i_t}(\eta^{t-1}, \xi^{t-1}) \subset \Theta_t \times \Xi_t. \tag{4.28}$$

Notice that Θ_{t,J,i_t} and Ξ_{t,J,i_t} can be viewed as measurable multifunctions since the vertices are measurable single-valued functions. This is a direct consequence of [91, proposition 1H]. Apart from measurability of the vertices and the inclusion (4.28), we require

$$\operatorname{supp} P_{t,J,i_t}(\cdot|\eta^{t-1}, \xi^{t-1}) \subset \Theta_{t,J,i_t}(\eta^{t-1}, \xi^{t-1}) \times \Xi_{t,J,i_t}(\eta^{t-1}, \xi^{t-1}). \tag{4.29}$$

A partition of the form (4.25) with the properties (4.26)–(4.29) always exists. In fact, the regularity condition (B4) requires the probability distribution Q_t of the disturbances $(\varepsilon_t^{\mathrm{o}}, \varepsilon_t^{\mathrm{r}})$ at stage t to be independent of the outcome history. Thus, for every time index $t \in \tau_{-0}$ and for every refinement parameter $J \in \mathbb{N}$ the measure Q_t can be represented as a convex combination of $I_t(J)$ regular probability measures with a (potentially) reduced support (see Sect. 4.3).

$$Q_t = \sum_{i_t=1}^{I_t(J)} \hat{\varrho}_{t,J,i_t} Q_{t,J,i_t}, \quad \sum_{i_t=1}^{I_t(J)} \hat{\varrho}_{t,J,i_t} = 1 \quad \forall t \in \tau_{-0},\, J \in \mathbb{N} \tag{4.30}$$

Here, Q_{t,J,i_t} is a regular probability measure, while $\hat{\varrho}_{t,J,i_t}$ is a nonnegative real number for every admissible combination of t, J, and i_t. Any partition of Q_t as in (4.30) naturally induces a partition of the conditional probability P_t as in (4.25). Concretely speaking, given the partition (4.30) we may define the transition probability P_{t,J,i_t} through

$$\begin{aligned} &P_{t,J,i_t}(A|\boldsymbol{\eta}^{t-1}, \boldsymbol{\xi}^{t-1}) \\ &= Q_{t,J,i_t}(\{(\varepsilon_t^{\mathrm{o}}, \varepsilon_t^{\mathrm{r}}) \,|\, (H_t^{\mathrm{o}}\boldsymbol{\eta}^{t-1} + \varepsilon_t^{\mathrm{o}}, H_t^{\mathrm{r}}\boldsymbol{\xi}^{t-1} + \varepsilon_t^{\mathrm{r}}) \in A\}), \end{aligned} \tag{4.31}$$

where A is an arbitrary Borel subset of $\Theta_t \times \Xi_t$, and $(\boldsymbol{\eta}^{t-1}, \boldsymbol{\xi}^{t-1})$ is an arbitrary element of the marginal space $\Theta^{t-1} \times \Xi^{t-1}$. Next, define ρ_{t,J,i_t} as a constant functional equal to the nonnegative real number $\hat{\rho}_{t,J,i_t}$. Moreover, for every admissible t, J, and i_t let $\mathcal{E}_{t,J,i_t}^{\mathrm{o}} \times \mathcal{E}_{t,J,i_t}^{\mathrm{r}}$ be a regular cross-simplex contained in $\mathcal{E}^{\mathrm{o}} \times \mathcal{E}^{\mathrm{r}}$ which covers the support of the probability measure Q_{t,J,i_t}. Then, the multifunctions (4.27) can be defined as

$$\begin{aligned} \Theta_{t,J,i_t}(\boldsymbol{\eta}^{t-1}, \boldsymbol{\xi}^{t-1}) &:= \{H_t^{\mathrm{o}}\boldsymbol{\eta}^{t-1}\} + \mathcal{E}_{t,J,i_t}^{\mathrm{o}}, \\ \Xi_{t,J,i_t}(\boldsymbol{\eta}^{t-1}, \boldsymbol{\xi}^{t-1}) &:= \{H_t^{\mathrm{r}}\boldsymbol{\xi}^{t-1}\} + \mathcal{E}_{t,J,i_t}^{\mathrm{r}}. \end{aligned} \tag{4.32}$$

Obviously, the vertices of these simplices reduce to linear affine and, a fortiori, measurable functions of the outcome history, and the inclusions (4.28) and (4.29) are satisfied by construction.

Given any partition (4.25) of the conditional probability P_t, we then attempt to discretize the associated transition probabilities P_{t,J,i_t} separately by means of the barycentric approximations. For ease of exposition, we suppress the fixed indices J and i_t in the subsequent discussion and merely consider possible dependencies on time and outcome history.

For any outcome history $(\boldsymbol{\eta}^{t-1}, \boldsymbol{\xi}^{t-1})$, the stage t outcomes $\boldsymbol{\eta}_t$ and $\boldsymbol{\xi}_t$ can be represented as convex combinations of the vertices in (4.27). Thus, the corresponding coordinates, i.e. the classical *barycentric weights*, depend measurably on the outcome history, cf. also (4.3).

$$\begin{aligned} \lambda_t(\boldsymbol{\eta}_t|\boldsymbol{\eta}^{t-1}, \boldsymbol{\xi}^{t-1}) &= (\lambda_{t,0}(\boldsymbol{\eta}_t|\boldsymbol{\eta}^{t-1}, \boldsymbol{\xi}^{t-1}), \ldots, \lambda_{t,K_t}(\boldsymbol{\eta}_t|\boldsymbol{\eta}^{t-1}, \boldsymbol{\xi}^{t-1})) \\ \tau_t(\boldsymbol{\xi}_t|\boldsymbol{\eta}^{t-1}, \boldsymbol{\xi}^{t-1}) &= (\tau_{t,0}(\boldsymbol{\xi}_t|\boldsymbol{\eta}^{t-1}, \boldsymbol{\xi}^{t-1}), \ldots, \tau_{t,L_t}(\boldsymbol{\xi}_t|\boldsymbol{\eta}^{t-1}, \boldsymbol{\xi}^{t-1})) \end{aligned}$$

Obviously, measurability is inherited by the *generalized barycentric weights.*

$$\gamma_{t,\nu,\mu}(\eta_t,\xi_t|\eta^{t-1},\xi^{t-1}) = \lambda_{t,\nu}(\eta_t|\eta^{t-1},\xi^{t-1})\,\tau_{t,\mu}(\xi_t|\eta^{t-1},\xi^{t-1})$$

Having in mind the extremal probability measures constructed in Sects. 4.2.3 and 4.2.4, we can now define two sequences of *pseudo-probabilities.*

$$q_{\Xi_t}(\mu_t|\eta^{t-1},\xi^{t-1}) := \int \tau_{\mu_t}(\xi_t|\eta^{t-1},\xi^{t-1})\,dP_t(\eta_t,\xi_t|\eta^{t-1},\xi^{t-1}) \quad (4.33a)$$

$$q_{\Theta_t}(\nu_t|\eta^{t-1},\xi^{t-1}) := \int \lambda_{\nu_t}(\eta_t|\eta^{t-1},\xi^{t-1})\,dP_t(\eta_t,\xi_t|\eta^{t-1},\xi^{t-1}) \quad (4.33b)$$

The pseudo-probabilities are well-defined, strictly positive, and measurable in the outcome history as implied by the measurability of the classical barycentric weights, the properties of the transition probability P_t, and the generalized Fubini theorem [2, theorem 2.6.4]. In the spirit of the previous sections, we define two sequences of *generalized barycenters* as

$$\eta_{\mu_t}(\eta^{t-1},\xi^{t-1}) := \sum_{\nu_t=0}^{K_t} a_{\nu_t}(\eta^{t-1},\xi^{t-1}) \times \int \frac{\gamma_{\nu_t,\mu_t}(\eta_t,\xi_t|\eta^{t-1},\xi^{t-1})}{q_{\Xi_t}(\mu_t|\eta^{t-1},\xi^{t-1})}\,dP_t(\eta_t,\xi_t|\eta^{t-1},\xi^{t-1}) \quad (4.34a)$$

and

$$\xi_{\nu_t}(\eta^{t-1},\xi^{t-1}) := \sum_{\mu_t=0}^{L_t} b_{\mu_t}(\eta^{t-1},\xi^{t-1}) \times \int \frac{\gamma_{\nu_t,\mu_t}(\eta_t,\xi_t|\eta^{t-1},\xi^{t-1})}{q_{\Theta_t}(\nu_t|\eta^{t-1},\xi^{t-1})}\,dP_t(\eta_t,\xi_t|\eta^{t-1},\xi^{t-1}). \quad (4.34b)$$

By means of elementary arguments it can be shown that the generalized barycenters are well-defined and measurable in the outcome history. Next, we use the pseudo-probabilities (4.33a) and the generalized barycenters (4.34a) to construct a parametric family of probability measures $P_t^l(\cdot|\eta^{t-1},\xi^{t-1})$, each of which has a finite support

$$\{(\eta_{\mu_t}(\eta^{t-1},\xi^{t-1}), b_{\mu_t}(\eta^{t-1},\xi^{t-1}))\,|\,\mu_t = 0,\ldots,L_t\}$$

and associated probabilities $q_{\Xi_t}(\mu_t|\eta^{t-1},\xi^{t-1})$. Notice that P_t^l characterizes a discrete transition probability approximating the original transition probability P_t. Analogously, we construct a transition probability P_t^u corresponding to the pseudo-probabilities (4.33b) and the generalized barycenters (4.34b). In this case, the discrete probability measure $P_t^u(\cdot|\eta^{t-1},\xi^{t-1})$ associated with the outcome history (η^{t-1},ξ^{t-1}) has support

$$\{(a_{\nu_t}(\eta^{t-1},\xi^{t-1}), \xi_{\nu_t}(\eta^{t-1},\xi^{t-1}))\,|\,\nu_t = 0,\ldots,K_t\},$$

and the ν_t'th atom has mass $q_{\Theta_t}(\nu_t|\eta^{t-1},\xi^{t-1})$. Reintroducing the suppressed indices J and i_t, we denote by P^l_{t,J,i_t} and P^u_{t,J,i_t} the barycentric discretizations of the transition probability P_{t,J,i_t}. Then, convex combination yields discrete approximates $P^l_{t,J}$ and $P^u_{t,J}$ for the original conditional probability P_t.

$$P^l_{t,J}(\cdot|\eta^{t-1},\xi^{t-1}) := \sum_{i_t=1}^{I_t(J)} \varrho_{t,J,i_t}(\eta^{t-1},\xi^{t-1}) P^l_{t,J,i_t}(\cdot|\eta^{t-1},\xi^{t-1}) \quad (4.35a)$$

$$P^u_{t,J}(\cdot|\eta^{t-1},\xi^{t-1}) := \sum_{i_t=1}^{I_t(J)} \varrho_{t,J,i_t}(\eta^{t-1},\xi^{t-1}) P^u_{t,J,i_t}(\cdot|\eta^{t-1},\xi^{t-1}) \quad (4.35b)$$

Measurability of the coefficient functions guarantees that both $P^l_{t,J}$ and $P^u_{t,J}$ can be interpreted as transition probabilities, too. One may improve the approximations of the conditional probability P_t by partitioning some or all components P_{t,J,i_t} of some initial partition (4.25) indexed by the given parameter J. The nested partition, which will be indexed by $J+1$, is referred to as a *refinement* of partition J. By successively refining the partition of P_t, thereby increasing the number $I_t(J)$ of components, one can construct two sequences of discrete *barycentric transition probabilities* $\{P^l_{t,J}\}_{J\in\mathbb{N}}$ and $\{P^u_{t,J}\}_{J\in\mathbb{N}}$ approximating the original conditional probability P_t.

By the product measure theorem [2, Sect. 2.6], the transition probabilities $P^l_{t,J}$ and $P^u_{t,J}$ of stages $t \in \tau_{-0}$ and the degenerate marginal probability measures $P^l_{0,J} := P_0 =: P^u_{0,J}$ can be combined to a unique 'lower' and 'upper' Borel probability measure P^l_J and P^u_J, respectively (see also the general procedure described in Sect. 4.1).

$$P^l_J := P^l_{0,J} * P^l_{1,J} * \cdots * P^l_{T,J}, \quad P^u_J := P^u_{0,J} * P^u_{1,J} * \cdots * P^u_{T,J} \quad (4.36)$$

Below, we will refer to P^l_J and P^u_J as *barycentric probability measures.* Observe that both barycentric measures have a finite support. Their atoms as well as the associated probabilities can be calculated by means of a forward recursion scheme, as exemplified in [43]. Furthermore, the transition probabilities (4.35) can be viewed as *regular* conditional probabilities corresponding to the barycentric measures (4.36). By convention, $E^l_{t,J}(\cdot)$ and $E^u_{t,J}(\cdot)$ stands for expectation under the barycentric measures conditional on information available at time t, while the unconditional expectation operators are denoted by $E^l_J(\cdot)$ and $E^u_J(\cdot)$, respectively.

A priori, for a fixed refinement parameter $J \in \mathbb{N}$, neither P^l_J nor P^u_J need be good substitutes for the original measure P in a given stochastic program. Occasionally, one might wish to improve the barycentric measures by refining the underlying transition probabilities. Thus, we will investigate the convergence behavior of the sequences $\{P^l_J\}_{J\in\mathbb{N}}$ and $\{P^u_J\}_{J\in\mathbb{N}}$, below. Under nonrestrictive conditions both sequences can be shown to converge weakly to the original measure P as the parameter J tends to infinity.

The underlying refinement strategy is arbitrary to a large extent. We will refer to a refinement strategy as *regular* if the following condition holds: for every tolerance $\varepsilon > 0$ there is a $J_0(\varepsilon) \in \mathbb{N}$ such that

$$\begin{gathered} \operatorname{diam}\left(\Theta_{t,J,i_t}(\boldsymbol{\eta}^{t-1},\boldsymbol{\xi}^{t-1}) \times \Xi_{t,J,i_t}(\boldsymbol{\eta}^{t-1},\boldsymbol{\xi}^{t-1})\right) \leq \varepsilon \\ \forall\, t \in \tau_{-0},\, J \geq J_0(\varepsilon),\, i_t = 1,\ldots,I_t(J),\, (\boldsymbol{\eta}^{t-1},\boldsymbol{\xi}^{t-1}) \in \Theta^{t-1} \times \Xi^{t-1}. \end{gathered} \tag{4.37}$$

This condition is nonrestrictive. For instance, concentrating on partitions of the form (4.30)–(4.32), one can easily construct a refinement strategy which satisfies (4.37). Generally speaking, the regularity condition (4.37) ensures weak convergence of $\{P^l_J\}_{J\in\mathbb{N}}$ and $\{P^u_J\}_{J\in\mathbb{N}}$ to the original probability measure P, as argued in the following proposition.

Proposition 4.4. *For any regular refinement strategy the sequences of discrete barycentric measures $\{P^l_J\}_{J\in\mathbb{N}}$ and $\{P^u_J\}_{J\in\mathbb{N}}$ on the augmented probability space converge weakly to the original probability measure P.*

Proof. Consider a continuous function $\Phi : \Theta \times \Xi \to \mathbb{R}$, and define

$$q_t := \begin{cases} (E_t\Phi) & \text{for } t < T, \\ \quad \Phi & \text{for } t = T. \end{cases}$$

The conditions (B1) and (B4) imply via the dominated convergence theorem that q_t is uniformly continuous on $\Theta^t \times \Xi^t$ for all $t \in \tau$. Thus, for every tolerance $\varepsilon > 0$ there is an index $J_0(\varepsilon) \in \mathbb{N}$ such that

$$|(E_t q_{t+1}) - (E^l_{t,J} q_{t+1})| \leq \varepsilon \tag{4.38}$$

uniformly on the marginal space $\Theta^t \times \Xi^t$ for all $t \in \tau_{-T}$ and for all sufficiently large refinement parameters $J \geq J_0(\varepsilon)$. In fact, proposition 4.3 guarantees that (4.38) holds pointwise, and uniformity of the estimate follows from uniformity of the assumption (4.37). For the sake of transparent notation, in the rest of the proof we suppress the index $J \geq J_0(\varepsilon)$ which will be kept fixed. We will show by backward induction that

$$|(E_t\Phi) - (E^l_t\Phi)| \leq \varepsilon\,(T-t)$$

uniformly on $\Theta^t \times \Xi^t$ for all $t \in \tau_{-T}$. The basis step for $t = T-1$ follows trivially from (4.38). Next, assume that the claim has been established for stage $t+1$. Thus, we find

$$\begin{aligned} |(E_t\Phi) - (E^l_t\Phi)| &= |(E_t(E_{t+1}\Phi)) - (E^l_t(E^l_{t+1}\Phi))| \\ &\leq |(E_t(E_{t+1}\Phi)) - (E^l_t(E_{t+1}\Phi))| + \varepsilon\,(T-t-1) \\ &\leq \varepsilon\,(T-t). \end{aligned}$$

The first inequality is due to the induction hypothesis, while the second inequality follows from (4.38). As the choice of ε was arbitrary, we have convergence of the unconditional expectation $(E^l\Phi)$ to $(E\Phi)$ due to refinements.

In turn, as the choice of the continuous function Φ was arbitrary, as well, weak convergence of the sequence of lower measures is established. An analog argument applies to the sequence of upper measures. □

By replacing the conditional probability P_t in the recourse problem (3.2) with the transition probabilities $P^l_{t,J}$ and $P^u_{t,J}$, a related *lower* and *upper auxiliary recourse problem* can be formulated, respectively. For each fixed refinement parameter $J \in \mathbb{N}$ we introduce a sequence of *auxiliary value functions* $\{\Phi^l_{t,J}\}_{t\in\tau}$ corresponding to the *lower* transition probabilities $P^l_{t,J}$ and an analogous sequence of *auxiliary value functions* $\{\Phi^u_{t,J}\}_{t\in\tau}$ corresponding to the *upper* transition probabilities $P^u_{t,J}$. The auxiliary value functions are defined by backward recursion. In the last decision stage set

$$\begin{aligned}\Phi^l_{T,J}(\boldsymbol{x}^{T-1},\boldsymbol{\eta}^T,\boldsymbol{\xi}^T) &:= \sup\nolimits_{\boldsymbol{x}_T} \hat{\rho}_T(\boldsymbol{x}^T,\boldsymbol{\eta}^T,\boldsymbol{\xi}^T),\\ \Phi^u_{T,J}(\boldsymbol{x}^{T-1},\boldsymbol{\eta}^T,\boldsymbol{\xi}^T) &:= \sup\nolimits_{\boldsymbol{x}_T} \hat{\rho}_T(\boldsymbol{x}^T,\boldsymbol{\eta}^T,\boldsymbol{\xi}^T),\end{aligned} \tag{4.39a}$$

and for $t = T-1,\ldots,1$ define

$$\begin{aligned}\Phi^l_{t,J}(\boldsymbol{x}^{t-1},\boldsymbol{\eta}^t,\boldsymbol{\xi}^t) &:= \sup\nolimits_{\boldsymbol{x}_t} \hat{\rho}_t(\boldsymbol{x}^t,\boldsymbol{\eta}^t,\boldsymbol{\xi}^t) + \big(E^l_{t,J}\Phi^l_{t+1,J}\big)(\boldsymbol{x}^t,\boldsymbol{\eta}^t,\boldsymbol{\xi}^t),\\ \Phi^u_{t,J}(\boldsymbol{x}^{t-1},\boldsymbol{\eta}^t,\boldsymbol{\xi}^t) &:= \sup\nolimits_{\boldsymbol{x}_t} \hat{\rho}_t(\boldsymbol{x}^t,\boldsymbol{\eta}^t,\boldsymbol{\xi}^t) + \big(E^u_{t,J}\Phi^u_{t+1,J}\big)(\boldsymbol{x}^t,\boldsymbol{\eta}^t,\boldsymbol{\xi}^t).\end{aligned} \tag{4.39b}$$

Finally, the auxiliary value functions of the subordinate first stage problems only depend on the random vector (η_0, ξ_0).

$$\begin{aligned}\Phi^l_{0,J}(\eta_0,\xi_0) &:= \sup\nolimits_{\boldsymbol{x}_0} \hat{\rho}_0(\boldsymbol{x}_0,\eta_0,\xi_0) + \big(E^l_{0,J}\Phi^l_{1,J}\big)(\boldsymbol{x}_0,\eta_0,\xi_0)\\ \Phi^u_{0,J}(\eta_0,\xi_0) &:= \sup\nolimits_{\boldsymbol{x}_0} \hat{\rho}_0(\boldsymbol{x}_0,\eta_0,\xi_0) + \big(E^u_{0,J}\Phi^u_{1,J}\big)(\boldsymbol{x}_0,\eta_0,\xi_0)\end{aligned} \tag{4.39c}$$

The expectation functionals are defined in the obvious way as in (3.3); simply replace the original conditional probabilities by the upper and lower barycentric transition probabilities, respectively. Furthermore, the optimal values

$$(E^l_J\Phi^l_{0,J}) := \int_{\Theta_0\times\Xi_0} \Phi^l_{0,J}(\eta_0,\xi_0)dP^l_{0,J}(\eta_0,\xi_0),$$

$$(E^u_J\Phi^u_{0,J}) := \int_{\Theta_0\times\Xi_0} \Phi^u_{0,J}(\eta_0,\xi_0)dP^u_{0,J}(\eta_0,\xi_0)$$

corresponding to the dynamic programs (4.39) are each given by the unconditional expectation of the respective first stage recourse function. It is intuitively clear that the barycentric probability measures P^l_J and P^u_J can be used to establish two *auxiliary stochastic programs*

$$\sup_{\boldsymbol{x}\in\mathcal{N}^l_{n,J}} \int_{\Theta\times\Xi} \left[\sum_{t=0}^{T} \rho_t(\boldsymbol{x}^t,\boldsymbol{\eta}^t)\right] dP^l_J(\boldsymbol{\eta},\boldsymbol{\xi}) \tag{4.40a}$$

$$\text{s.t.} \quad \boldsymbol{f}_t(\boldsymbol{x}^t,\boldsymbol{\xi}^t) \le \mathbf{0} \quad P^l_J\text{-a.s.} \quad t\in\tau$$

and

$$\sup_{\boldsymbol{x}\in\mathcal{N}^u_{n,J}} \int_{\Theta\times\Xi} \left[\sum_{t=0}^{T} \rho_t(\boldsymbol{x}^t,\boldsymbol{\eta}^t)\right] dP^u_J(\boldsymbol{\eta},\boldsymbol{\xi}) \tag{4.40b}$$
$$\text{s.t.}\quad \boldsymbol{f}_t(\boldsymbol{x}^t,\boldsymbol{\xi}^t) \leq \boldsymbol{0} \quad P^u_J\text{-a.s.} \quad t\in\tau$$

The linear spaces $\mathcal{N}^l_{n,J}$ and $\mathcal{N}^u_{n,J}$ contain all non-anticipative policy functions which are essentially bounded with respect to the measures P^l_J and P^u_J, respectively. We will refer to (4.40) as the *lower* and *upper auxiliary stochastic programs* associated with the underlying optimization problem (3.1). As the measures P^l_J and P^u_J are discrete, the auxiliary stochastic programs (4.40) are principally accessible for numerical solution.

4.5 Bounds on the Optimal Value

By successively applying the inequalities (4.17) and (4.21) to minorize (majorize) the expectation functionals of a stochastic program subject to the regularity conditions (B1)–(B5), it can be shown that the optimal value is bounded below by $(E^l_J\Phi^l_{0,J})$ and bounded above by $(E^u_J\Phi^u_{0,J})$. Little surprisingly, these bounds become tendentially tighter as the refinement parameter J is increased. Unlike the true optimal value, the bounds are computationally accessible, as their calculation merely requires the solution of finite-dimensional stochastic programs. Moreover, the difference $(E^u_J\Phi^u_{0,J}) - (E^l_J\Phi^l_{0,J})$ can be viewed as a measure for the 'appropriateness' of the auxiliary stochastic programs (4.40) to approximate the original problem (3.1). These ideas will be formalized in the remainder of this section.

Proposition 4.5. *Consider a stochastic program of the form (3.1), which satisfies the fundamental regularity conditions (B1)–(B5). Then, the following hold for all $J\in\mathbb{N}$:*

(a) the transition probabilities $P^l_{t,J}$ and $P^u_{t,J}$ are well-defined for $t\in\tau$ (but not unique) and define two discrete probability measures P^l_J and P^u_J on the measurable space $(\Theta\times\Xi,\mathcal{B}(\Theta\times\Xi))$;

(b) the auxiliary stochastic programs (4.40) satisfy the regularity conditions (A1)"–(A5)";

(c) the auxiliary value functions $\Phi^l_{t,J}$ and $\Phi^u_{t,J}$ are concave in the decision vector $\boldsymbol{x}^{t-1}$ for all $t\in\tau_{-0}$.

Proof. Assertion (a) follows from (B1), (B4), the construction of the transition probabilities $P^l_{t,J}$ and $P^u_{t,J}$, as well as the product measure theorem [2, Sect. 2.6]. Assertion (b) is straightforward. By construction, the auxiliary stochastic programs (4.40) are subject to the conditions (B1), (B2), (B3), and (B5). However, they do not necessarily satisfy condition (B4). Thus, arguing as in corollary 3.15, one shows that the auxiliary value functions $\Phi^l_{t,J}$ and $\Phi^u_{t,J}$ are concave in the decision variables on the entire underlying space. This observation implies assertion (c). Due to failure of assumption (B4), subdifferentiability as well as the saddle property can not be established. □

Proposition 4.5 (b) implies that the static and dynamic versions of the auxiliary stochastic programs (4.40) are solvable and equivalent. Furthermore, the auxiliary recourse functions are usc and bounded on their natural domains. In view of the results in proposition 4.1 one might expect monotonicity of the auxiliary recourse functions upon increase of the refinement parameter. This is in fact provable provided that the auxiliary stochastic programs (4.40) comply with the regularity condition (B4); see proposition 4.6 below. However – as remarked in the proof of proposition 4.5 – condition (B4) usually fails to be true. Consequently, monotonicity fails to hold in general.

Proposition 4.6 (Monotonicity). *Consider two sequences of lower and upper auxiliary stochastic programs of the form (4.40), each of which satisfies the fundamental regularity conditions (B1)–(B5). Then, for all $t \in \tau$ we have*

$$\Phi^l_{t,J} \geq \Phi^l_{t,J'} \quad \textit{and} \quad \Phi^u_{t,J} \leq \Phi^u_{t,J'} \quad \textit{on } Z^t \quad \forall J \geq J'. \tag{4.41}$$

Proof. The inequalities (4.41) are shown by backward induction with respect to the index t. By (4.39a) we have $\Phi^l_{T,J} = \Phi^l_{T,J'}$ and $\Phi^u_{T,J} = \Phi^u_{T,J'}$ for all refinement parameters $J, J' \in \mathbb{N}$. Thus, the basis step is obvious. Then, assume that (4.41) has already been established for stage $t+1$. As the auxiliary stochastic programs are subject to the conditions (B1)–(B5), we may conclude by the theorems 3.14 and 3.18 that $\Phi^l_{t,J}$ and $\Phi^u_{t,J}$ are subdifferentiable saddle functions on Z^t for all $(t, J) \in \tau \times \mathbb{N}$. Thus, for all $J \geq J'$ we find

$$\begin{aligned}(E^l_{t,J}\Phi^l_{t+1,J}) \geq (E^l_{t,J}\Phi^l_{t+1,J'}) \geq (E^l_{t,J'}\Phi^l_{t+1,J'})\\ (E^u_{t,J}\Phi^u_{t+1,J}) \leq (E^u_{t,J}\Phi^u_{t+1,J'}) \leq (E^u_{t,J'}\Phi^u_{t+1,J'})\end{aligned}$$

on the natural domain Y^t. The first inequalities on both lines are due to the induction hypothesis, while the second inequalities follow from proposition 4.1 as well as the saddle property of $\Phi^l_{t+1,J'}$ and $\Phi^u_{t+1,J'}$, respectively. An elementary argument then proves (4.41) for stage t, which completes the induction step. □

The assumptions of proposition 4.6 are fulfilled e.g. if the underlying original stochastic program complies with the regularity conditions (B1)–(B5), and for each $(t, J) \in \tau_{-0} \times \mathbb{N}$ the J'th partition of P_t is of the form (4.30)–(4.32).

Below, the crucial theorem 4.7 clarifies under what conditions the auxiliary stochastic programs (4.40) provide lower and upper bounds on the optimal value of the original stochastic program (3.1).

Theorem 4.7 (Bounds on the Recourse Functions). *Consider a multi-stage stochastic program, which satisfies the conditions (B1)–(B5). Then, Φ_t is squeezed between the auxiliary value functions, i.e. we have $\Phi^l_{t,J} \leq \Phi_t \leq \Phi^u_{t,J}$ on Z^t for all $(t, J) \in \tau \times \mathbb{N}$. Similarly, the optimal value of the original problem is bounded:*

$$(E^l_J \Phi^l_{0,J}) \leq (E\Phi_0) \leq (E^u_J \Phi^u_{0,J}).$$

Proof. For the sake of better readability we may suppress the index J, which is kept fixed in this proof. Theorem 3.14 and theorem 3.18 imply that the recourse functions $\{\Phi_t\}_{t\in\tau}$ are subdifferentiable saddle functions on a neighborhood of their natural domains. This notion is important for the present theorem, which will be proved by backward induction. By definition we have

$$\Phi^l_T(x^{T-1}, \eta^T, \xi^T) = \Phi_T(x^{T-1}, \eta^T, \xi^T) = \Phi^u_T(x^{T-1}, \eta^T, \xi^T),$$

and thus the basis step is established. Let us now assume that we have already shown $\Phi^l_{t+1} \leq \Phi_{t+1} \leq \Phi^u_{t+1}$ on Z^{t+1}. By using the barycentric approximations we can easily derive a similar chain of inequalities for the expectation functionals.

$$\begin{aligned}
&\int \Phi^l_{t+1}(x^t, \eta^{t+1}, \xi^{t+1})\, dP^l_{t+1}(\eta_{t+1}, \xi_{t+1}|\eta^t, \xi^t) \\
\leq &\int \Phi_{t+1}(x^t, \eta^{t+1}, \xi^{t+1})\, dP^l_{t+1}(\eta_{t+1}, \xi_{t+1}|\eta^t, \xi^t) \\
\leq &\int \Phi_{t+1}(x^t, \eta^{t+1}, \xi^{t+1})\, dP_{t+1}(\eta_{t+1}, \xi_{t+1}|\eta^t, \xi^t) \\
\leq &\int \Phi_{t+1}(x^t, \eta^{t+1}, \xi^{t+1})\, dP^u_{t+1}(\eta_{t+1}, \xi_{t+1}|\eta^t, \xi^t) \\
\leq &\int \Phi^u_{t+1}(x^t, \eta^{t+1}, \xi^{t+1})\, dP^u_{t+1}(\eta_{t+1}, \xi_{t+1}|\eta^t, \xi^t)
\end{aligned}$$

The first and the fourth inequalities follow from the induction hypothesis, whereas the second and the third inequalities apply because of (4.17), (4.21), and the construction of the transition probabilities $P^{l,u}_{t+1}$; make also use of the saddle structure and the subdifferentiability of Φ_{t+1}. Comparison of the objective functions in (3.2b) and (4.39b) entails

$$\Phi^l_t(x^{t-1}, \eta^t, \xi^t) \leq \Phi_t(x^{t-1}, \eta^t, \xi^t) \leq \Phi^u_t(x^{t-1}, \eta^t, \xi^t).$$

This statement completes the induction step. □

Theorem 4.7 is a combination of lemma 4.2 and theorem 4.1 in [43]. Subsequently, we investigate the convergence properties of the auxiliary value functions (4.39). As a key ingredient we will need condition (4.37), which basically ensures weak convergence of the sequences $\{P^l_J\}_{J\in\mathbb{N}}$ and $\{P^u_J\}_{J\in\mathbb{N}}$ to the original probability measure P. The following theorem is motivated by the results in [44, Sect. 5].

Theorem 4.8 (Convergence). *Consider a multistage stochastic program subject to the regularity conditions (B1)–(B5). Then, the sequences* $\{\Phi^l_{t,J}\}_{J\in\mathbb{N}}$ *and* $\{\Phi^u_{t,J}\}_{J\in\mathbb{N}}$ *converge to* Φ_t *uniformly on* Z^t *for all* $t\in\tau$ *provided that the underlying refinement strategy is regular (see (4.37)).*

Proof. By theorem 3.18 the recourse function Φ_t is uniformly continuous on its natural domain Z^t for all $t\in\tau$. Thus, for every tolerance $\varepsilon>0$ we find

$$(E_t\Phi_{t+1})-(E^l_{t,J}\Phi_{t+1})\leq\varepsilon \tag{4.42}$$

uniformly on Y^t for all $t\in\tau_{-T}$ and for all sufficiently large refinement parameters $J\geq J_0(\varepsilon)$. In fact, proposition 4.3 guarantees that (4.42) holds pointwise, and uniformity of the estimate follows from uniformity of the assumption (4.37). For the sake of transparency, in the rest of the proof we suppress the index $J\geq J_0(\varepsilon)$, which will be kept fixed. We will argue that

$$0\leq\Phi_t-\Phi^l_t\leq\varepsilon\,(T-t)$$

uniformly on Z^t for all $t\in\tau$. The claim is established by backward induction. The basis step is trivially fulfilled since, by definition, Φ_T equals Φ^l_T on the entire underlying space. Next, assume that $\Phi_{t+1}-\Phi^l_{t+1}\leq\varepsilon\,(T-t-1)$. Then, we may conclude that

$$\begin{aligned}\Phi^l_t(\boldsymbol{x}^{t-1},\boldsymbol{\eta}^t,\boldsymbol{\xi}^t)&=\sup_{\boldsymbol{x}_t}\hat{\rho}_t(\boldsymbol{x}^t,\boldsymbol{\eta}^t,\boldsymbol{\xi}^t)+(E^l_t\Phi^l_{t+1})(\boldsymbol{x}^t,\boldsymbol{\eta}^t,\boldsymbol{\xi}^t)\\&\geq\sup_{\boldsymbol{x}_t}\hat{\rho}_t(\boldsymbol{x}^t,\boldsymbol{\eta}^t,\boldsymbol{\xi}^t)+(E^l_t\Phi_{t+1})(\boldsymbol{x}^t,\boldsymbol{\eta}^t,\boldsymbol{\xi}^t)-\varepsilon\,(T-t-1)\\&\geq\sup_{\boldsymbol{x}_t}\hat{\rho}_t(\boldsymbol{x}^t,\boldsymbol{\eta}^t,\boldsymbol{\xi}^t)+(E_t\Phi_{t+1})(\boldsymbol{x}^t,\boldsymbol{\eta}^t,\boldsymbol{\xi}^t)-\varepsilon\,(T-t)\\&=\Phi_t(\boldsymbol{x}^{t-1},\boldsymbol{\eta}^t,\boldsymbol{\xi}^t)-\varepsilon\,(T-t)\end{aligned}$$

uniformly on Z^t. The first inequality is due to the induction hypothesis, while the second inequality follows from (4.42). Furthermore, by theorem 4.7 we find $\Phi_t-\Phi^l_t\geq 0$ uniformly on Z^t, which completes the induction step. As the tolerance ε was arbitrary, convergence of the lower sequence $\{\Phi^l_{t,J}\}_{J\in\mathbb{N}}$ to the original recourse function Φ_t is established. An analog argument applies to the upper sequence $\{\Phi^u_{t,J}\}_{J\in\mathbb{N}}$. □

4.6 Bounding Sets for the Optimal Decisions

In the previous section we developed computable bounds bracketing the optimal value of a stochastic program which satisfies the regularity conditions

(B1)–(B5). In practice, however, a decision maker is not only interested in the optimal value but also in the optimal policy of a given stochastic optimization problem. In the sequel, we will argue that bounds on the recourse functions entail *bounding sets* for the optimal first stage decisions. Our exposition is inspired by Pflug [84, Sect. 1.3].[6] As before, it is convenient to use the standard assumption that the random outcomes at stage 0 are deterministic. Thus, one may consider $\boldsymbol{\eta}_0$ and $\boldsymbol{\xi}_0$ as fixed parameters implying that

$$(E\Phi_0) = \Phi_0(\boldsymbol{\eta}_0, \boldsymbol{\xi}_0), \quad (E^l_J\Phi^l_0) = \Phi^l_{0,J}(\boldsymbol{\eta}_0, \boldsymbol{\xi}_0), \quad (E^u_J\Phi^u_0) = \Phi^u_{0,J}(\boldsymbol{\eta}_0, \boldsymbol{\xi}_0).$$

Next, introduce three extended-real-valued functionals

$$\begin{aligned} F(\boldsymbol{x}_0) &:= \hat{\rho}_0(\boldsymbol{x}_0, \boldsymbol{\eta}_0, \boldsymbol{\xi}_0) + (E\ \Phi_1)(\boldsymbol{x}_0, \boldsymbol{\eta}_0, \boldsymbol{\xi}_0), \\ F^l_J(\boldsymbol{x}_0) &:= \hat{\rho}_0(\boldsymbol{x}_0, \boldsymbol{\eta}_0, \boldsymbol{\xi}_0) + (E^l_J\Phi^l_{1,J})(\boldsymbol{x}_0, \boldsymbol{\eta}_0, \boldsymbol{\xi}_0), \\ F^u_J(\boldsymbol{x}_0) &:= \hat{\rho}_0(\boldsymbol{x}_0, \boldsymbol{\eta}_0, \boldsymbol{\xi}_0) + (E^u_J\Phi^u_{1,J})(\boldsymbol{x}_0, \boldsymbol{\eta}_0, \boldsymbol{\xi}_0). \end{aligned} \tag{4.43}$$

We assume that the auxiliary functionals F^l_J and F^u_J can numerically be evaluated, whereas F is computationally untractable. Moreover, assume the underlying stochastic program to satisfy the conditions (B1)–(B5). Then we have

$$\begin{aligned} (E\Phi_0) &= \sup_{\boldsymbol{x}_0} F(\boldsymbol{x}_0), \\ (E^l_J\Phi^l_{0,J}) &= \sup_{\boldsymbol{x}_0} F^l_J(\boldsymbol{x}_0), \\ (E^u_J\Phi^u_{0,J}) &= \sup_{\boldsymbol{x}_0} F^u_J(\boldsymbol{x}_0), \end{aligned}$$

and $F^l_J \leq F \leq F^u_J$ on $\mathbb{R}^{n_0}$. Therefore, the functional F is squeezed between the auxiliary functionals F^l_J and F^u_J on the entire underlying space. This observation implies that the maximizers of F are necessarily contained in the set of all $\boldsymbol{x}_0$ such that $F^u_J(\boldsymbol{x}_0)$ is greater or equal to the maximum of the lower bounding function F^l_J (see Fig. 4.3). Formally, one may write

$$\arg\max F \subset C_J := \{\boldsymbol{x}_0 \,|\, F^u_J(\boldsymbol{x}_0) \geq \sup F^l_J\}. \tag{4.44}$$

It is obvious from the definitions that the upper level set C_J also contains the maximizers of the auxiliary functionals (see Fig. 4.3), i.e.

$$\arg\max F^l_J \cup \arg\max F^u_J \subset C_J. \tag{4.45}$$

Moreover, implementation of any first stage decision in $\arg\max F^l_J$ results in an expected profit of at least $\sup F^l$, provided that all recourse decisions

[6]In that reference, three lsc functions F, F^l, and F^u are considered. These functions are subject to specific conditions. First, F is squeezed between F^l and F^u, i.e. $F^l \leq F \leq F^u$. Moreover, the minima of F^l and F^u are finite and known, and F^l satisfies a suitable growth condition. Then, the distance between the minimizers of F and F^l can be estimated.

are chosen optimally with respect to the original probability measure. Notice that C_J represents a bounding set for the optimal decisions at stage 0 as it contains all maximizers of F. Moreover, C_J is compact. This follows from the observation that C_J is an upper level set of F_J^u and thus contained in the first stage feasible set $X_0(\boldsymbol{\xi}_0)$. The level set C_J can principally be found by scanning the space of first stage decisions. In fact, one can always decide whether some $\boldsymbol{x}_0 \in \mathbb{R}^{n_0}$ lies within C^J by simply calculating $F_J^u(\boldsymbol{x}_0)$ and comparing the result with $\sup F_J^l$. Thus, scanning requires maximization of F_J^l and repeated evaluation of F_J^u at specific reference points, both of which are practicable operations. However, unsystematic scanning is very inefficient for $n_0 > 1$. In case of a multidimensional decision space one should exploit structural properties of the functionals (4.43). By assumption, the underlying stochastic program complies with the regularity conditions (B1)–(B5). Then, proposition 4.5 ensures concavity of F_J^l and F_J^u. In particular, $F_J^u(\boldsymbol{x}_0)$ can be interpreted as the optimal value of (4.40b) given that the first stage decision is fixed at $\boldsymbol{x}_0$. Any Lagrange multiplier associated with this extra constraint in (4.40b) – i.e. the equality constraint which fixes $\boldsymbol{x}_0$ – represents a subgradient of F_J^u at $\boldsymbol{x}_0$. Subgradient information of this type might be useful for an efficient numerical evaluation of the level set C_J.

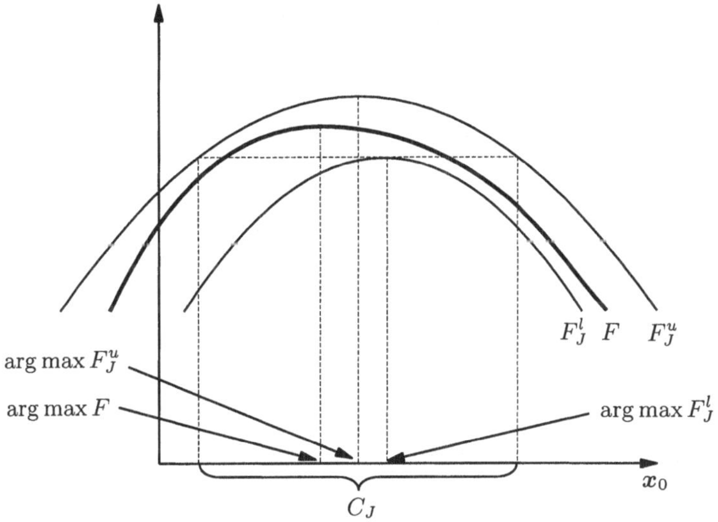

Fig. 4.3. Bounding set for the optimal first stage decisions

The above analysis can be further simplified by concentrating on one single component of the first stage decision vector. Denote by $F_{J,k}^l(x_{0,k})$ and $F_{J,k}^u(x_{0,k})$ the optimal values of the auxiliary stochastic programs (4.40a) and (4.40b), respectively, and let $F_k(x_{0,k})$ be the optimal value of (3.1) under the additional assumption that the kth component of $\boldsymbol{x}_0$ is fixed at $x_{0,k}$. By definition, we find

$$\begin{aligned}
F_k(x_{0,k}) &= \sup\{F(\boldsymbol{x}_0') \mid x_{0,k}' = x_{0,k}\}, \\
F_{J,k}^l(x_{0,k}) &= \sup\{F_J^l(\boldsymbol{x}_0') \mid x_{0,k}' = x_{0,k}\}, \\
F_{J,k}^u(x_{0,k}) &= \sup\{F_J^u(\boldsymbol{x}_0') \mid x_{0,k}' = x_{0,k}\},
\end{aligned}$$

and $F_{J,k}^l \leq F_k \leq F_{J,k}^u$ on $\mathbb{R}$ for all $k = 1, \ldots, n_0$. Moreover, we have

$$\begin{aligned}
(E\Phi_0) &= \sup_{x_{0,k}} F_k(x_{0,k}), \\
(E_J^l \Phi_{0,J}^l) &= \sup_{x_{0,k}} F_{J,k}^l(x_{0,k}), \\
(E_J^u \Phi_{0,J}^u) &= \sup_{x_{0,k}} F_{J,k}^u(x_{0,k}).
\end{aligned}$$

The maximizers of F_k can be estimated in analogy to (4.44).

$$\arg\max F_k \subset C_{J,k} := \{\boldsymbol{x}_0 \mid F_{J,k}^u(\boldsymbol{x}_0) \geq \sup F_{J,k}^l\} \tag{4.46}$$

However, (4.46) has the advantage that $F_{J,k}^l$ and $F_{J,k}^u$ are functions of one single real variable only.[7] Upper and lower bounds on the optimal $x_{0,k}$ are thus provided by the two zeros of the concave function $F_{J,k}^u - \sup F_{J,k}^l$, which can conveniently be calculated by means of Newton-type methods.

Now, let us return to the multidimensional bounding sets C_J introduced in (4.44). As argued in the following theorem, the sequence $\{C_J\}_{J \in \mathbb{N}}$ converges[8] to the set of maximizers of the functional F.

Theorem 4.9 (Convergence). *Consider a multistage stochastic program subject to the regularity conditions (B1)–(B5). Then, the sequence $\{C_J\}_{J \in \mathbb{N}}$ converges to $\arg\max F$ provided that the underlying refinement strategy is regular in the sense of (4.37).*

Proof. By definition of the bounding set C_J it is clear that

$$\arg\max F \subset C_J \subset \operatorname{dom} F \quad \forall J \in \mathbb{N}, \tag{4.47}$$

where $\operatorname{dom} F$ coincides with the first stage feasible set $X_0(\boldsymbol{\xi}_0)$. In addition, we will prove the following implication:

$$\begin{aligned}
\boldsymbol{x}_0 \notin \arg\max F \Rightarrow & \text{ there is a neighborhood } U \text{ of } \boldsymbol{x}_0 \text{ and} \\
& J_0 \in \mathbb{N} \text{ such that } U \cap C_J = \emptyset \ \forall J \geq J_0.
\end{aligned} \tag{4.48}$$

---

[7] Notice, however, that the intervals $\{C_{J,k}\}_{k=1}^{n_0}$ do not provide full information about the bounding set C_J. In contrast, C_J is usually a strict subset of the Cartesian product $\times_{k=1}^{n_0} C_{J,k}$.

[8] For a survey on the theory of set convergence see [97, Chap. 4].

In fact, given a fixed decision $\boldsymbol{x}_0$ in the complement of $\arg\max F$, we will show that some neighborhood U of $\boldsymbol{x}_0$ lies in the complement of C_J for all sufficiently large refinement parameters J. As $\boldsymbol{x}_0$ is no optimizer of F, and by closedness of $\arg\max F$, there is a compact neighborhood U of $\boldsymbol{x}_0$ such that

$$U \cap \arg\max F = \emptyset.$$

Upper semicontinuity of F and compactness of U then ensure the existence of an $\varepsilon > 0$ with

$$F(\boldsymbol{x}_0') < \sup F - 2\varepsilon \quad \forall \boldsymbol{x}_0' \in U \cap \operatorname{dom} F.$$

Moreover, by theorem 4.8 the sequences $\{F_J^l\}_{J\in\mathbb{N}}$ and $\{F_J^u\}_{J\in\mathbb{N}}$ converge to the functional F uniformly on $\operatorname{dom} F$. Consequently, there exists a refinement parameter $J_0 \in \mathbb{N}$ such that

$$F_J^u - F < \varepsilon \text{ and } F - F_J^l < \varepsilon$$

on $\operatorname{dom} F$ and for all $J \geq J_0$. Combining the above inequalities we obtain

$$F_J^u(\boldsymbol{x}_0') < F(\boldsymbol{x}_0') + \varepsilon < \sup F - \varepsilon < \sup F_J^l \quad \forall \boldsymbol{x}_0' \in U \cap \operatorname{dom} F.$$

Then, by the definition (4.44) of the bounding set C_J, it is immediately clear that the set U lies in the complement of C_J for all $J \geq J_0$. Thus, (4.48) follows. In summary, we may conclude that

$$\arg\max F \subset \liminf_{J\to\infty} C_J \subset \limsup_{J\to\infty} C_J \subset \arg\max F,$$

where the first inclusion follows from (4.47), and the third inclusion follows from (4.48). Thus, the limit of the sequence $\{C_J\}_{J\in\mathbb{N}}$ exists and coincides with the set of maximizers of F. This observation completes the proof. □

Corollary 4.10. *Let $\{\boldsymbol{x}_{0,J}^\star\}_{J\in\mathbb{N}}$ be a converging sequence in $\mathbb{R}^{n_0}$ with limiting point $\boldsymbol{x}_0^\star$. Moreover, assume that $\boldsymbol{x}_{0,J}^\star \in \arg\max F_J^l$ for every $J \in \mathbb{N}$. Then, $\boldsymbol{x}_0^\star$ is an element of $\arg\max F$. The same holds true if we assume that $\boldsymbol{x}_{0,J}^\star \in \arg\max F_J^u$ for every $J \in \mathbb{N}$.*

Proof. The assertion follows immediately from the inclusion (4.45) and the fact that the bounding sets $\{C_J\}_{J\in\mathbb{N}}$ converge to $\arg\max F$. □

In [42, Sect. 18] the statement of corollary 4.10 is derived for the two-stage case by using the concept of *epi-convergence* due to Attouch and Wets [3]. The approach presented here provides sharper results, as we know for sure that $\arg\max F$ is covered by the compact bounding set C_J for every $J \in \mathbb{N}$, and the sequence $\{C_J\}_{J\in\mathbb{N}}$ converges to $\arg\max F$.

5

Extensions

Many decision problems under uncertainty, which are of economic or technical interest, can be formulated as multistage stochastic programs. As we have seen in the previous section, such optimization problems are conveniently discretized by means of the barycentric approximation scheme. Loosely speaking, the most crucial prerequisites for barycentric approximation to yield upper and lower bounds on the recourse functions are:

(a) the given stochastic program is a convex optimization problem (without loss of generality, we concentrate on maximization problems);

(b) the profit functions are convex in the stochastic parameters;

(c) the constraint functions are jointly convex in the decision variables and the stochastic parameters;

(d) the stochastic parameters follow a block-diagonal autoregressive process.

Although the first requirement seems to be fairly restrictive, many optimization problems of practical relevance are actually convex. However, the remaining conditions (b)–(d) severely limit the scope of the barycentric approximation scheme.[1] In an economic context, for instance, decision problems involving lognormally distributed prices and demands, derivative trading, or risk-aversion do not meet the requirements (b), (c), and (d) simultaneously. Sometimes, these conditions can be enforced by redefining the random parameters and appropriately transforming the underlying probability distributions.[2] But if the random variables are serially correlated, this approach generally fails. From an applied point of view it would be advantageous to circumvent the unnatural restrictions (b) and (c) and to extend the barycentric approximation

[1]Naive application of the barycentric approximation scheme to problems which fail to satisfy the conditions (b)–(d) may provide useful results, but no error bounds.

[2]This is always possible for linear two-stage stochastic programs.

scheme to broader problem classes. However, from a theoretical perspective it is highly questionable whether the auxiliary value functions then still provide upper and lower bounds on the original recourse functions. In order to clarify the role of the auxiliary value functions in a less restrictive setting, we will study below the structural properties of convex multistage stochastic programs with a general nonconvex dependence on the random variables. In particular, Sect. 5.1 is devoted to the study of profit functions which are nonconvex in $\boldsymbol{\eta}$, whereas Sect. 5.2 investigates generalized constraint functions which are nononvex in $\boldsymbol{\xi}$. We will argue that under weak regularity conditions the recourse functions of these generalized stochastic programs still have arbitrarily tight upper and lower bounds, which are computationally accessible. Moreover, the bounds basically coincide with the well-known auxiliary value functions shifted by specific random variables.

Below, we frequently work with the barycentric measures (4.36). Moreover, we further investigate the auxiliary recourse functions (4.39) introduced in Chap. 4. However, for the sake of transparent notation we will usually suppress the refinement parameter $J \in \mathbb{N}$.

5.1 Stochasticity of the Profit Functions

In this section we consider again a convex multistage stochastic program of the form (3.1). As argued in Sect. 3.5, the assumptions (B1)–(B5) ensure that the recourse function Φ_t of such an optimization problem is subdifferentiable and has a characteristic saddle structure on a neighborhood of Z^t, $t \in \tau$. In the sequel, we shall modify the technical condition (B2) in order to allow for profit functions with a nonconvex $\boldsymbol{\eta}$-dependence. To this end, we need the following definition.

Definition 5.1. *For any $t \in \tau$ the profit function ρ_t is called regularizable if there exists a continuous mapping $\alpha_t : \mathbb{R}^{K^t} \to \mathbb{R}$ with the following properties: α_t is convex in $\boldsymbol{\eta}^t$ on a convex neighborhood of Θ^t, and $\rho_t + \alpha_t$ is a saddle function on a convex neighborhood of Y^t being concave in $\boldsymbol{x}^t$, convex in $\boldsymbol{\eta}^t$, and constant in $\boldsymbol{\xi}^t$.*

Notice that if ρ_t is regularizable, then it is necessarily concave in $\boldsymbol{x}^t$, constant in $\boldsymbol{\xi}^t$, and continuous on a convex neighborhood of Y^t. Below we will refer to the functions $\{\alpha_t\}_{t\in\tau}$ as *correction terms*. Next, we can state the modified regularity conditions.

(C1) = (B1);
(C2) – the profit function ρ_t is continuous on the underlying Euclidean space and concave in $\boldsymbol{x}^t$ for fixed values of the stochastic parameters, $t \in \tau$;
– ρ_t is regularizable for every $t \in \tau$;
(C3) = (B3); (C4) = (B4); (C5) = (B5).

These conditions are assumed to hold throughout this section. Note that assumption (C2) includes assumption (B2) as a special case. In fact, if (B2) holds, the trivial correction terms $\{\alpha_t \equiv 0\}_{t\in\tau}$ suffice to regularize the profit functions of all decision stages. However, it is important to realize that regularizable profit functions can have a generalized nonconvex $\boldsymbol{\eta}$-dependence. Below, proposition 5.3 identifies a large class of regularizable profit functions and demonstrates the construction of suitable correction terms.

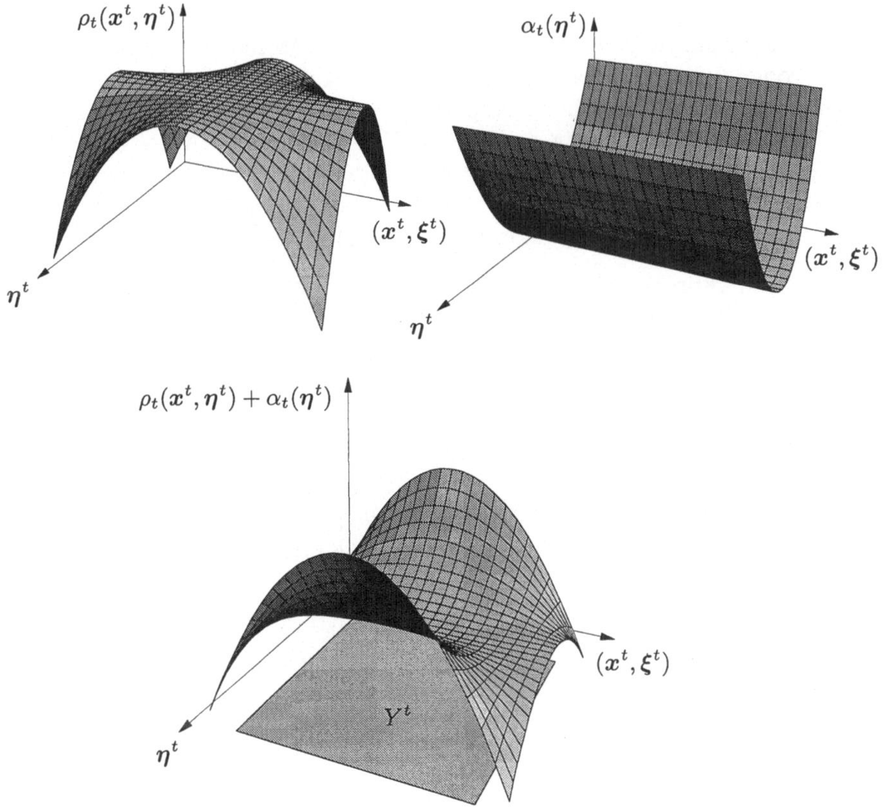

Fig. 5.1. Regularization of a biconcave profit function

Suppressing the refinement parameter $J \in \mathbb{N}$ for better readability, it is useful to introduce the auxiliary value functions $\{\Phi_t^l\}_{t\in\mathcal{T}}$ and $\{\Phi_t^u\}_{t\in\mathcal{T}}$ as in (4.39) and the discrete barycentric measures P^l and P^u as in (4.36). The following proposition 5.2 ensures that the auxiliary value functions and the recourse functions of the original problem are well-defined.

Proposition 5.2. *Consider a multistage stochastic program satisfying the regularity conditions (C1)–(C5). Then the following hold:*

(a) the transition probabilities P_t^l and P_t^u are well-defined for $t \in \tau$ and define two discrete probability measures P^l and P^u on $(\Theta \times \Xi, \mathcal{B}(\Theta \times \Xi))$;

(b) the original stochastic program complies with the conditions (A1)–(A5), and the associated auxiliary programs satisfy (A1)"–(A5)";

(c) the value functions Φ_t, Φ_t^l, and Φ_t^u are concave in the decision vector $\boldsymbol{x}^{t-1}$ for all $t \in \tau_{-0}$.

Proof. Assertion (a) follows from (C1) and (C4). Moreover, the propositions 3.6 through 3.9 remain valid under the new regularity conditions (C1)–(C5) without modification of the corresponding proofs. These results can be used to show that the regularity conditions (C1)–(C5) entail (A1)–(A5). As the barycentric measures are discrete, we may then conclude that the auxiliary stochastic programs satisfy (A1)"–(A5)". Thus, (b) follows. Finally, statement (c) is proved precisely as in proposition 4.5. □

Part (b) of proposition 5.2 implies that the static and dynamic versions of the stochastic program under consideration are both solvable and equivalent. The same is true for the associated auxiliary stochastic programs. Moreover, the generalized feasible sets Y^t and Z^t are convex and compact for each $t \in \tau$. Compactness is needed for the proof of the following proposition 5.3, which basically states that smooth profit functions are always regularizable.

Proposition 5.3. *For some $t \in \tau$ assume that ρ_t is a smooth function on a convex neighborhood of the compact set Y^t being twice continuously differentiable in $\boldsymbol{\eta}^t$, concave in $\boldsymbol{x}^t$, and constant in $\boldsymbol{\xi}^t$. Then, ρ_t is regularizable.*

Proof. By compactness of Y^t there is a compact convex neighborhood U of Y^t and an open convex neighborhood V of U such that the restriction of ρ_t to V is twice continuously differentiable in $\boldsymbol{\eta}^t$, i.e. the second order partial derivatives of ρ_t with respect to the stochastic variables are continuous on V. Consider the assignment which maps any vector in V to the Hessian submatrix corresponding to the components of $\boldsymbol{\eta}^t$.

$$(\boldsymbol{x}^t, \boldsymbol{\eta}^t, \boldsymbol{\xi}^t) \mapsto \nabla_{\boldsymbol{\eta}^t} \otimes \nabla_{\boldsymbol{\eta}^t}\, \rho_t(\boldsymbol{x}^t, \boldsymbol{\eta}^t) \in \mathbb{R}^{K^t \times K^t}$$

By assumption, this matrix-valued function is continuous on V. Compactness of $U \subset V$ and continuity of the matrix 2-norm (denoted by $\|\cdot\|_2$) imply that

$$c_t^2 := \sup\left\{ \left\|\nabla_{\boldsymbol{\eta}^t} \otimes \nabla_{\boldsymbol{\eta}^t} \rho_t(\boldsymbol{x}^t, \boldsymbol{\eta}^t)\right\|_2 \,\middle|\, (\boldsymbol{x}^t, \boldsymbol{\eta}^t, \boldsymbol{\xi}^t) \in U) \right\}$$

is finite and nonnegative. Thus, the regularized matrix $\nabla_{\boldsymbol{\eta}^t} \otimes \nabla_{\boldsymbol{\eta}^t} \rho_t + c_t^2\, 1_{K^t}$ is positive semidefinite on U, where 1_{K^t} denotes the K^t-dimensional identity matrix. Moreover, the function $\rho_t(\boldsymbol{x}^t, \boldsymbol{\eta}^t) + c_t^2 \langle \boldsymbol{\eta}^t, \boldsymbol{\eta}^t \rangle$ is convex in $\boldsymbol{\eta}^t$ on U. Based upon these notions we can define

$$\alpha_t(\boldsymbol{\eta}^t) := c_t^2 \langle \boldsymbol{\eta}^t, \boldsymbol{\eta}^t \rangle.$$

By construction, $\rho_t + \alpha_t$ is a continuous saddle function on a convex neighborhood of Y^t. Thus, ρ_t is regularizable. □

In the remainder of this section we consider a stochastic program of the form (3.1) subject to the regularity conditions (C1)–(C5). For every $t \in \tau$, α_t denotes a suitable correction term as in definition 5.1, which is assumed to be given. It should be emphasized again that these correction terms are required to be continuous and constant in the decision variables. In order to simplify notation, let us introduce three sequences of additional random variables. For every $t \in \tau$ define

$$\begin{aligned} A_t(\boldsymbol{\eta}^t) &:= E_t\left[\sum_{s=t}^{T} \alpha_s(\boldsymbol{\eta}^s)\right], \\ A_t^l(\boldsymbol{\eta}^t) &:= E_t^l\left[\sum_{s=t}^{T} \alpha_s(\boldsymbol{\eta}^s)\right], \\ A_t^u(\boldsymbol{\eta}^t) &:= E_t^u\left[\sum_{s=t}^{T} \alpha_s(\boldsymbol{\eta}^s)\right]. \end{aligned} \tag{5.1}$$

Thus, A_t, A_t^l, and A_t^u are given by the conditional expectation values of the future correction terms with respect to the probability measures P, P^l, and P^u, respectively. In addition, we will need the unconditional expectation of the sum of all correction terms with respect to the three probability measures under consideration:

$$\begin{aligned} (E\, A_0) &:= E\ \left[A_0(\boldsymbol{\eta}_0)\right], \\ (E^l A_0^l) &:= E^l \left[A_0^l(\boldsymbol{\eta}_0)\right], \\ (E^u A_0^u) &:= E^u \left[A_0^u(\boldsymbol{\eta}_0)\right]. \end{aligned} \tag{5.2}$$

The random variables (5.1) as well as the constants (5.2) will be referred to as *conditional correction terms* below. It is important to notice that the conditional correction terms are computationally accessible. Evaluation of A_t^l and A_t^u requires calculation of a finite sum, while A_t can be evaluated by means of numerical integration techniques. Since α_t is continuous for every $t \in \tau$, A_t, A_t^l, and A_t^u are continuous and bounded on Θ^t. Moreover, by convexity of the correction terms we find $A_t^l \leq A_t \leq A_t^u$ on Θ^t, although A_t^l and A_t^u are not necessarily convex (use a similar argument as in the proof of theorem 4.7). With the above conventions we are now prepared to define an *adjusted recourse problem*

$$\begin{aligned} \sup_{\boldsymbol{x} \in \mathcal{N}_n} \int_{\Theta \times \Xi} & \left[\sum_{t=0}^{T} \rho_t(\boldsymbol{x}^t, \boldsymbol{\eta}^t) + \alpha_t(\boldsymbol{\eta}^t) \right] dP(\boldsymbol{\eta}, \boldsymbol{\xi}) \\ \text{s.t.} \quad & \boldsymbol{f}_t(\boldsymbol{x}^t, \boldsymbol{\xi}^t) \leq \mathbf{0} \quad P\text{-a.s.} \quad t \in \tau, \end{aligned} \tag{5.3}$$

where the stage t profit function is given by $\rho_t + \alpha_t$. Let Ψ_t be the optimal value function of stage t associated with the stochastic program (5.3). As the correction terms $\{\alpha_t\}_{t \in \tau}$ are independent of the decision variables, they have no influence on the optimal policy and only cause a shift in the recourse functions and the optimal objective function value. Moreover, denote by Ψ_t^l and Ψ_t^u the auxiliary value functions of stage t corresponding to the recourse problem (5.3). The following proposition 5.4 summarizes some important properties of the adjusted recourse problem and the associated auxiliary value functions.

Proposition 5.4. *Consider a multistage stochastic program subject to the conditions (C1)–(C5), and define the adjusted recourse problem as in (5.3). Then, we may conclude:*

(a) $\Psi_t^l \leq \Psi_t \leq \Psi_t^u$ *on* Z^t $\forall t \in \tau$*;*

(b) $(E^l \Psi_0^l) \leq (E \Psi_0) \leq (E^u \Psi_0^u)$*;*

(c) $\Psi_t = \Phi_t + A_t$ *and* $\Psi_t^d = \Phi_t^d + A_t^d$ $\forall d \in \{l, u\}$, $t \in \tau$*;*

(d) $(E\Psi_0) = (E\Phi_0) + (EA_0)$ *and* $(E^d \Psi_0^d) = (E^d \Phi_0^d) + (E^d A_0^d)$ $\forall d \in \{l, u\}$*.*

Proof. By construction, the profit functions $\{\rho_t + \alpha_t\}_{t \in \tau}$ satisfy the regularity condition (B2). Consequently, the adjusted recourse problem (5.3) satisfies each of the conditions (B1)–(B5). This observation implies via the theorems 3.14 and 3.18 that Ψ_t is subdifferentiable and exhibits a saddle structure on a convex neighborhood of Z^t ($t \in \tau$). In particular, the assumptions of theorem 4.7 are fulfilled, and the barycentric approximation scheme applies yielding upper and lower bounds on the recourse function Ψ_t. Hence the assertions (a) and (b) follow. The statements (c) and (d) are straightforward and follow directly from the definition of the adjusted recourse problem and the fact that the correction terms are independent of the decision variables. □

Theorem 5.5 (Adjusted Bounds I). *Consider a stochastic program satisfying the regularity conditions (C1)–(C5). Then, the correction terms* $\{\alpha_t\}_{t \in \tau}$ *can be used to construct (finite) bounds on the recourse function, i.e. we find*

$$\Phi_t^l + A_t^l - A_t \le \Phi_t \le \Phi_t^u + A_t^u - A_t \quad \textit{on } Z^t \textit{ for all } t \in \tau.$$

Analogously, the optimal objective function value has bounds:

$$(E^l \Phi_0^l) + (E^l A_0^l) - (EA_0) \le (E\Phi_0) \le (E^u \Phi_0^u) + (E^u A_0^u) - (EA_0).$$

Proof. The claim follows directly from proposition 5.4. □

Theorem 5.6 (Convergence). *Consider a stochastic program satisfying the regularity conditions (C1)-(C5). If the barycentric probability measures are regularly refined in the sense of (4.37), then the random variables A_t^l and A_t^u converge to A_t uniformly on Θ^t, while the auxiliary recourse functions Φ_t^l and Φ_t^u converge to Φ_t uniformly on Z^t for all $t \in \tau$.*

Proof. By assumption, the correction term α_t is uniformly continuous on Θ^t for every $t \in \tau$. Thus, convergence of A_t^l and A_t^u to A_t uniformly on Θ^t $(t \in \tau)$ follows from the weak convergence of the discrete barycentric measures to the original probability measure. Moreover, theorem 4.8 applied to the adjusted recourse problem (5.3) ensures convergence of Ψ_t^l and Ψ_t^u to Ψ_t uniformly on Z^t for all $t \in \tau$. Combining the above results, by proposition 5.4 (c) we have convergence of Φ_t^l and Φ_t^u to Φ_t uniformly on Z^t for all $t \in \tau$. □

5.2 Stochasticity of the Constraint Functions

This section is devoted to the study of generalized constraint functions. Concretely speaking, we aim at modifying assumption (B3) in order to allow for constraint functions with a generalized nonconvex $\boldsymbol{\xi}$-dependence. For the further argumentation we need the following definition.

Definition 5.7. *For any $t \in \tau$ the constraint function $\boldsymbol{f}_t$ is called regularizable if there is a continuous vector-valued mapping $\boldsymbol{\kappa}_t : \mathbb{R}^{L^t} \to \mathbb{R}^{r_t}$ with the following properties: $\boldsymbol{\kappa}_t$ is convex in $\boldsymbol{\xi}^t$ on a convex neighborhood of Ξ^t, while $\boldsymbol{f}_t + \boldsymbol{\kappa}_t$ is jointly convex in $(\boldsymbol{x}^t, \boldsymbol{\xi}^t)$ and constant in $\boldsymbol{\eta}^t$ on a convex neighborhood of Y^t.*

It is convenient to write the mapping $\boldsymbol{\kappa}_t$ as

$$\boldsymbol{\kappa}_t = (\boldsymbol{\kappa}_t^{\text{in}}, \boldsymbol{\kappa}_t^{\text{eq}+}, \boldsymbol{\kappa}_t^{\text{eq}-}) \quad \text{where} \quad \begin{cases} \boldsymbol{\kappa}_t^{\text{in}} \;\; : \mathbb{R}^{L^t} \to \mathbb{R}^{r_t^{\text{in}}}, \\ \boldsymbol{\kappa}_t^{\text{eq}+} : \mathbb{R}^{L^t} \to \mathbb{R}^{r_t^{\text{eq}}}, \\ \boldsymbol{\kappa}_t^{\text{eq}-} : \mathbb{R}^{L^t} \to \mathbb{R}^{r_t^{\text{eq}}}. \end{cases}$$

This representation reflects the grouping of $\boldsymbol{f}_t = (\boldsymbol{f}_t^{\text{in}}, \boldsymbol{f}_t^{\text{eq}}, -\boldsymbol{f}_t^{\text{eq}})$ by inequality and equality constraints. Notice that any regularizable constraint function

$\boldsymbol{f}_t$ is necessarily convex in $\boldsymbol{x}^t$, constant in $\boldsymbol{\eta}^t$, and locally Lipschitz[3] on a convex neighborhood of Y^t. Local Lipschitz continuity follows from the fact that $\boldsymbol{f}_t$ can be written as the difference of the functions $\boldsymbol{f}_t + \boldsymbol{\kappa}_t$ and $\boldsymbol{\kappa}_t$, which are both convex and locally Lipschitz on a convex neighborhood of Y^t. Below, we will prove compactness of Y^t, which allows us to conclude that $\boldsymbol{f}_t$ is even (globally) Lipschitz on some compact convex neighborhood of Y^t. But first, an important result about linear constraints is provided in the following proposition.

Proposition 5.8. *Consider a constraint function $\boldsymbol{f}_t$ which is linear affine in the decision variables, i.e.*

$$\boldsymbol{f}_t(\boldsymbol{x}^t, \boldsymbol{\xi}^t) = W_t(\boldsymbol{\xi}^t)\boldsymbol{x}_t + T_t(\boldsymbol{\xi}^t)\boldsymbol{x}^{t-1} - \boldsymbol{h}_t(\boldsymbol{\xi}^t), \tag{5.4}$$

where W_t, T_t, and $\boldsymbol{h}_t$ are matrix- and vector-valued mappings on $\mathbb{R}^{L^t}$ with appropriate dimensions. If $\boldsymbol{f}_t$ is regularizable, then W_t and T_t are constant and $\boldsymbol{h}_t$ is representable as a difference of two convex functions on a convex neighborhood of Ξ^t.

Proof. Since the constraint function $\boldsymbol{f}_t$ is regularizable, there is a vector-valued mapping $\boldsymbol{\kappa}_t$ as in definition 5.7. Recall that such a mapping is continuous on the entire space and convex on a convex neighborhood of Ξ^t. Applying proposition D.1 in appendix D to each component of $\boldsymbol{f}_t + \boldsymbol{\kappa}_t$ separately and for every reference point in Y^t, we may conclude that the matrices W_t and T_t are locally constant while the difference function $\boldsymbol{\kappa}_t - \boldsymbol{h}_t$ is locally convex on a convex neighborhood of Ξ^t. By connectedness of convex sets, W_t and T_t are constant on a convex neighborhood of Ξ^t. Convexity of $\boldsymbol{f}_t + \boldsymbol{\kappa}_t$ then entails convexity of $\boldsymbol{\kappa}_t - \boldsymbol{h}_t$ on a convex neighborhood of Ξ^t. Consequently, the mapping

$$\boldsymbol{h}_t = \boldsymbol{\kappa}_t - (\boldsymbol{\kappa}_t - \boldsymbol{h}_t)$$

is representable as a difference of two convex functions on a convex neighborhood of Ξ^t. This notion completes the proof. □

With obvious modifications, proposition 5.8 applies to the linear affine components of any regularizable constraint function. For example, consider a regularizable constraint function $\boldsymbol{f}_t = (\boldsymbol{f}_t^{\mathrm{in}}, \boldsymbol{f}_t^{\mathrm{eq}}, -\boldsymbol{f}_t^{\mathrm{eq}})$ which accounts for both equality and inequality constraints. Thus, both $\boldsymbol{f}_t^{\mathrm{eq}}$ as well as $-\boldsymbol{f}_t^{\mathrm{eq}}$ are convex in $\boldsymbol{x}^t$ for fixed values of the stochastic parameters. This implies that $\boldsymbol{f}_t^{\mathrm{eq}}$ is of the form (5.4), and the assertions of proposition 5.8 hold true. On the other hand, we know that a constraint function $\boldsymbol{f}_t$ which is linear affine in the decision variables can always be brought to the form (5.4). Then,

---

[3]Recall that if the single constraint function $f_{t,i}$ has Lipschitz constant $\Lambda_{t,i}$ for $i = 1, \ldots, r_t$, then $\boldsymbol{f}_t$ has Lipschitz constant $\Lambda_t := (\sum_{i=1}^{r_t} \Lambda_{t,i}^2)^{1/2}$.

proposition 5.8 implies that $\boldsymbol{f}_t$ is not regularizable if the coefficient matrices W_t or T_t depend nontrivially on the random parameters, i.e. if either W_t or T_t represents a nonconstant function of $\boldsymbol{\xi}^t$. This observation will be important for the study of linear stochastic programs in Sect. 5.4. Furthermore, in the remainder of this section, proposition 5.8 will be used to prove compactness of the generalized feasible sets $\{Y^t\}_{t\in\mathcal{T}}$ corresponding to a system of regularizable constraint functions.

Let us now study the properties of a multistage stochastic program of the form (3.1) subject to the following regularity conditions:

(D1)	=	(B1); (D2) = (B2);
(D3)	–	the constraint function $\boldsymbol{f}_t$ is continuous on the underlying Euclidean space and globally convex in $\boldsymbol{x}^t$ for fixed values of the stochastic parameters, $t \in \mathcal{T}$;
	–	$\boldsymbol{f}_t$ is regularizable for every $t \in \mathcal{T}$;
	–	the feasible set mapping X_t is bounded on a neighborhood of Z^t, $t \in \mathcal{T}$;
(D4)	=	(B4); (D5) = (B5).

These conditions are assumed to hold throughout this section. One should pay attention to a subtlety in the assumptions (D2) and (D3). In fact, these regularity conditions are required to hold on a convex neighborhood of Y^t, although Y^t itself is not necessarily convex. Thus, (D2) and (D3) must occasionally hold far beyond the natural domain of the profit and constraint functions (and not just on some ε-neighborhood of Y^t).

Note that condition (D3) includes condition (B3) as a special case. In fact, if (B3) holds, the trivial mappings $\{\boldsymbol{\kappa}_t \equiv \mathbf{0}\}_{t\in\mathcal{T}}$ suffice to regularize the constraint functions of all decision stages. Moreover, in proposition 5.11 we will present a broad class of regularizable constraint functions which do not satisfy assumption (B3).

Proposition 5.9. *Consider a stochastic program subject to the regularity conditions (D1)–(D5). Then, for all $t \in \mathcal{T}$ we have:*

(a) the feasible set mapping X_t is non-empty-compact-valued and usc on a neighborhood of Z^t;

(b) for any neighborhood U of Y^t there is a neighborhood V of Z^t such that

$$\operatorname{graph} X_t|_V \subset U.$$

Proof. By proposition 5.8, the restriction of $\boldsymbol{f}_t^{\mathrm{eq}}$ to some neighborhood of Y^t is representable as a sum of a linear function independent of $\boldsymbol{\xi}^t$ and a

Lipschitzian function independent of $\boldsymbol{x}^t$ for all $t \in \tau$. This result guarantees that the propositions 3.7 through 3.9 remain valid under the new regularity conditions (D1)–(D5) without modification of the corresponding proofs. □

Suppressing the refinement parameter $J \in \mathbb{N}$ to simplify notation, we introduce the auxiliary value functions $\{\Phi_t^l\}_{t\in\tau}$ and $\{\Phi_t^u\}_{t\in\tau}$ as in (4.39) and the discrete measures P^l and P^u as in (4.36). The following proposition 5.10 ensures that the original as well as the auxiliary value functions are well-defined.

Proposition 5.10. *Consider a multistage stochastic program satisfying the regularity conditions (D1)–(D5). Then the following hold:*

(a) the transition probabilities P_t^l and P_t^u are well-defined for $t \in \tau$ and define two discrete probability measures P^l and P^u on $(\Theta \times \Xi, \mathcal{B}(\Theta \times \Xi))$;

(b) the original stochastic program complies with the conditions (A1)–(A5), and the associated auxiliary programs satisfy (A1)"–(A5)";

(c) the value functions Φ_t, Φ_t^l, and Φ_t^u are concave in the decision vector $\boldsymbol{x}^{t-1}$ for all $t \in \tau_{-0}$.

Proof. Assertion (a) follows from (D1) and (D4). Arguing as in Sect. 3.3, one may use proposition 5.9 to show that the regularity conditions (D1)–(D5) entail the more fundamental conditions (A1)–(A5). As the barycentric measures are discrete, we may then conclude that the auxiliary stochastic programs satisfy (A1)"–(A5)". Thus, (b) follows. Finally, statement (c) is proved precisely as in proposition 4.5. □

Statement (b) of proposition 5.10 implies that the static and dynamic versions of the stochastic program under consideration are both solvable and equivalent. The same is true for the associated auxiliary stochastic programs. Moreover, the generalized feasible sets Y^t and Z^t are compact for each $t \in \tau$. Compactness of the sets $\{Y^t\}_{t\in\tau}$ helps us to identify an important class of regularizable constraint functions, as pointed out in the following proposition.

Proposition 5.11. *For some $t \in \tau$ assume that Y^t is compact, and the constraint function $\boldsymbol{f}_t$ is additively separable, i.e. $\boldsymbol{f}_t = \boldsymbol{f}_t^1 + \boldsymbol{f}_t^2$. Furthermore, we postulate*

(i) $\boldsymbol{f}_t^1$ is a continuous convex mapping on a convex neighborhood of Y^t being jointly convex in $(\boldsymbol{x}^t, \boldsymbol{\xi}^t)$ and constant in $\boldsymbol{\eta}^t$;

(ii) $\boldsymbol{f}_t^2$ is a smooth mapping on a convex neighborhood of Y^t being twice continuously differentiable in $\boldsymbol{\xi}^t$ and constant in $(\boldsymbol{x}^t, \boldsymbol{\eta}^t)$;

Then, $\boldsymbol{f}_t$ is regularizable.

Proof. By compactness of Y^t and assumption (ii) there is a compact convex neighborhood U of Y^t and an open convex neighborhood V of U such that the component $\boldsymbol{f}_t^2$ is twice continuously differentiable on V. Consider the assignment which maps any vector in V to the Hessian submatrix of $f_{t,i}^2$ corresponding to the stochastic variables $\boldsymbol{\xi}^t$ $(i = 1, \ldots .r_t)$.

$$(\boldsymbol{x}^t, \boldsymbol{\eta}^t, \boldsymbol{\xi}^t) \mapsto \nabla_{\boldsymbol{\xi}^t} \otimes \nabla_{\boldsymbol{\xi}^t} f_{t,i}^2(\boldsymbol{\xi}^t) \in \mathbb{R}^{L^t \times L^t}$$

By assumption, this matrix-valued function is continuous on V. Compactness of $U \subset V$ and continuity of the matrix 2-norm imply that

$$c_{t,i}^2 := \sup \left\{ \left\| \nabla_{\boldsymbol{\xi}^t} \otimes \nabla_{\boldsymbol{\xi}^t} f_{t,i}^2(\boldsymbol{\xi}^t) \right\|_2 \,\middle|\, (\boldsymbol{x}^t, \boldsymbol{\eta}^t, \boldsymbol{\xi}^t) \in U) \right\}$$

is finite and nonnegative. Thus, the regularized function $f_{t,i}^2(\boldsymbol{\xi}^t) + c_{t,i}^2 \langle \boldsymbol{\xi}^t, \boldsymbol{\xi}^t \rangle$ is convex in $\boldsymbol{\xi}^t$ on U. Based upon these notions we can define

$$\kappa_{t,i}(\boldsymbol{\xi}^t) := c_{t,i}^2 \langle \boldsymbol{\xi}^t, \boldsymbol{\xi}^t \rangle \quad \forall i = 1, \ldots, r_t, \qquad \boldsymbol{\kappa}_t := (\kappa_{t,1}, \ldots, \kappa_{t,r_t}).$$

By construction, $\boldsymbol{f}_t + \boldsymbol{\kappa}_t$ is a continuous convex function on a convex neighborhood of Y^t being jointly convex in $(\boldsymbol{x}^t, \boldsymbol{\xi}^t)$ and constant in $\boldsymbol{\eta}^t$. This implies that $\boldsymbol{f}_t$ is regularizable. □

In the proof of proposition 5.11 we make use of separability to show that the constraint function $\boldsymbol{f}_t$ is regularizable. Conversely, one can also show $\boldsymbol{f}_t$ to be regularizable if it is twice continuously differentiable in $(\boldsymbol{x}^t, \boldsymbol{\xi}^t)$, constant in $\boldsymbol{\eta}^t$, and strictly convex in $\boldsymbol{x}^t$ on a convex neighborhood of Y^t. Thus, loosely speaking, additive separability or strict convexity in the decision variables are both sufficient to imply regularizablility. Unfortunately, strict convexity is rarely seen in realistic problems, which are predominantly governed by linear constraints.

In the remainder, we consider a stochastic program of the form (3.1) subject to the new regularity conditions (D1)–(D5). For every $t \in \tau$, $\boldsymbol{\kappa}_t$ denotes a suitable mapping as in definition 5.7, which is assumed to be given.

Now, consider the dynamic version of the original stochastic program. The parametric stage t subproblem reads

$$\begin{aligned} \sup_{\boldsymbol{x}_t \in \mathbb{R}^{n_t}} \quad & q_t(\boldsymbol{x}^t, \boldsymbol{\eta}^t, \boldsymbol{\xi}^t) \\ \text{s.t.} \quad & \boldsymbol{f}_t^{\text{in}}(\boldsymbol{x}^t, \boldsymbol{\xi}^t) \leq \mathbf{0} \\ & \boldsymbol{f}_t^{\text{eq}}(\boldsymbol{x}^t, \boldsymbol{\xi}^t) = \mathbf{0}, \end{aligned} \tag{5.5}$$

where $q_t := \rho_t + (E_t \Phi_{t+1})$ for $t \in \tau_{-T}$, and $q_T := \rho_T$. Notice that this representation explicitly distinguishes inequality and equality constraints. Let

$$\boldsymbol{d}_t^{\text{in}*} = \boldsymbol{d}_t^{\text{in}*}(\boldsymbol{x}^{t-1}, \boldsymbol{\eta}^t, \boldsymbol{\xi}^t) \quad \text{and} \quad \boldsymbol{d}_t^{\text{eq}*} = \boldsymbol{d}_t^{\text{eq}*}(\boldsymbol{x}^{t-1}, \boldsymbol{\eta}^t, \boldsymbol{\xi}^t)$$

be any pair of history-dependent Lagrange multipliers associated with the inequality and equality constraints in (5.5), respectively. Notice that $(\boldsymbol{d}_t^{\text{in}*}, \boldsymbol{d}_t^{\text{eq}*})$ always exists, although not being unique in general; see appendix B for an introduction to Lagrangian duality. In order to allow for a uniform treatment of inequality and equality constraints, we assign to every pair $(\boldsymbol{d}_t^{\text{in}*}, \boldsymbol{d}_t^{\text{eq}*})$ an r_t-dimensional vector

$$\boldsymbol{d}_t^* := (\boldsymbol{d}_t^{\text{in}*}, [\boldsymbol{d}_t^{\text{eq}*}]^+, [\boldsymbol{d}_t^{\text{eq}*}]^-),$$

where

$$[\boldsymbol{d}^*]^\pm := \max\{\pm \boldsymbol{d}^*, \boldsymbol{0}\} \quad \forall \boldsymbol{d}^* \in \mathbb{R}^r,$$

and 'max' denotes the componentwise maximization operator. By construction, $\boldsymbol{d}_t^*$ lies in the nonnegative orthant of $\mathbb{R}^{r_t}$ and can formally be interpreted as a Lagrange multiplier corresponding to the unified constraint function $\boldsymbol{f}_t = (\boldsymbol{f}_t^{\text{in}}, \boldsymbol{f}_t^{\text{eq}}, -\boldsymbol{f}_t^{\text{eq}})$. Next, we introduce a vector $\boldsymbol{D}_t^* \in \mathbb{R}^{r_t}$ which is an upper bound for the multipliers $\boldsymbol{d}_t^*(\boldsymbol{x}^{t-1}, \boldsymbol{\eta}^t, \boldsymbol{\xi}^t)$ uniformly over all outcome and decision histories in some neighborhood of Z^t. At present, we do anticipate that such a vector can be found under the assumptions (D1)–(D5). A formal proof of its existence is postponed to proposition 5.12. For the further argumentation, it is convenient to introduce a non-anticipative stochastic process $\{\alpha_t\}_{t\in\tau}$, where the random variables

$$\alpha_t(\boldsymbol{\xi}^t) := -\langle \boldsymbol{D}_t^*, \boldsymbol{\kappa}_t(\boldsymbol{\xi}^t)\rangle \tag{5.6}$$

will be denoted as *correction terms*. As in the previous section, we try to argue that the adjusted recourse problem with profit functions $\{\rho_t + \alpha_t\}_{t\in\tau}$ has nice structural properties in view of the barycentric approximation scheme. Again, the correction terms $\{\alpha_t\}_{t\in\tau}$ are continuous and independent of the decision variables. In order to simplify notation, we introduce three sequences of additional random variables. For $t \in \tau$ we define

$$\begin{aligned} A_t(\boldsymbol{\xi}^t) &:= E_t\left[\sum_{s=t}^T \alpha_s(\boldsymbol{\xi}^s)\right], \\ A_t^l(\boldsymbol{\xi}^t) &:= E_t^l\left[\sum_{s=t}^T \alpha_s(\boldsymbol{\xi}^s)\right], \\ A_t^u(\boldsymbol{\xi}^t) &:= E_t^u\left[\sum_{s=t}^T \alpha_s(\boldsymbol{\xi}^s)\right]. \end{aligned} \tag{5.7}$$

Thus, A_t, A_t^l, and A_t^u are given by the conditional expectation of the future correction terms with respect to the probability measures P, P^l, and P^u, respectively. Finally, we set

$$\begin{aligned} (E\ A_0) &:= E\ \left[A_0(\boldsymbol{\xi}_0)\right], \\ (E^l A_0^l) &:= E^l\left[A_0^l(\boldsymbol{\xi}_0)\right], \\ (E^u A_0^u) &:= E^u\left[A_0^u(\boldsymbol{\xi}_0)\right]. \end{aligned} \tag{5.8}$$

The random variables (5.7) as well as the constants (5.8) will be referred to as *conditional correction terms* below. It is important to notice that the conditional correction terms are computationally accessible. Evaluation of A_t^l and A_t^u requires calculation of a finite sum while A_t can be evaluated by means of numerical integration techniques. Since α_t is continuous for every $t \in \tau$, A_t, A_t^l, and A_t^u are continuous and bounded on Ξ^t. Moreover, by concavity of the correction terms we find $A_t^l \leq A_t \leq A_t^u$ on Ξ^t, although A_t^l and A_t^u are not necessarily concave (use a similar argument as in the proof of theorem 4.7). As inspired by the previous section we define the *adjusted recourse problem* as

$$\sup_{\boldsymbol{x} \in \mathcal{N}_n} \int_{\Theta \times \Xi} \left[\sum_{t=0}^{T} \rho_t(\boldsymbol{x}^t, \boldsymbol{\eta}^t) + \alpha_t(\boldsymbol{\xi}^t) \right] dP(\boldsymbol{\eta}, \boldsymbol{\xi}) \tag{5.9}$$
$$\text{s.t.} \quad \boldsymbol{f}_t(\boldsymbol{x}^t, \boldsymbol{\xi}^t) \leq \boldsymbol{0} \quad P\text{-a.s.} \quad t \in \tau.$$

Let $\Psi_t \equiv \Phi_t + A_t$ be the optimal value function of stage t associated with the stochastic program (5.9). As the correction terms are independent of the decision variables, they have no influence on the optimal policy and only cause a shift in the recourse functions and the optimal objective function value. Moreover, denote by $\Psi_t^l = \Phi_t^l + A_t^l$ and $\Psi_t^u = \Phi_t^u + A_t^u$ the lower and upper auxiliary value functions of stage t, respectively, corresponding to the adjusted recourse problem (5.9).

In the remainder of this section we will argue that the optimal value functions of the adjusted recourse problem (5.9) are subdifferentiable and exhibit a specific saddle structure. This requires more sophisticated arguments than in Sect. 5.1 as the adjusted recourse problem (5.9) generally fails to comply with the regularity conditions (B1)–(B5).

Proposition 5.12. *Consider a stochastic program, which satisfies the regularity conditions (D1)–(D5), and define the adjusted recourse problem as in (5.9) with recourse functions $\{\Psi_t\}_{t \in \tau}$. Then, we find for all $t \in \tau$:*

(a) there exists an upper bounding vector $\boldsymbol{D}_t^*$ as in the definition of the correction terms (5.6);

(b) Ψ_t is a subdifferentiable saddle function on a neighborhood of Z^t being convex in $\boldsymbol{\eta}^t$ and jointly concave in $\boldsymbol{\xi}^t$ and the decision variables.

Assertion (b) states that for each $t \in \tau$ the adjusted value function Ψ_t is a convex-concave saddle function on some neighborhood of its natural domain Z^t, which is generally nonconvex. Recall that the notion of convexity (concavity) for functions on nonconvex domains has been introduced in Sect. 3.2.2.

Proof of proposition 5.12. For each $t \in \tau$ we will construct an extended-real-valued auxiliary function $\bar{\Psi}_t$ which coincides with the recourse function Ψ_t

on a neighborhood of Z^t. The mapping $\bar{\Psi}_t$ is a saddle function being convex in $\boldsymbol{\eta}^t$ and jointly concave in $\boldsymbol{\xi}^t$ and the decision variables on the entire underlying space. Moreover it is subdifferentiable and continuous on a convex neighborhood of Z^t. As a byproduct, we can recursively show the existence of a bounding vector $\boldsymbol{D}_t^*$ as in the definition of the correction terms (5.6).

The claim is shown by backward induction. By the assumptions (D2) and (D3) there are open (not necessarily convex) sets

$$Y_{\cap}^T \subset \mathbb{R}^{n^T} \times \mathbb{R}^{L^T} \quad \text{and} \quad Y_{\cup}^T \subset \mathbb{R}^{K^T}$$

such that

(i) $Y_{\cap}^T \times Y_{\cup}^T$ is a neighborhood of Y^T;

(ii) ρ_T is a continuous saddle function on $\operatorname{co} Y_{\cap}^T \times \operatorname{co} Y_{\cup}^T$ being concave in $\boldsymbol{x}^T$, convex in $\boldsymbol{\eta}^T$, and constant in $\boldsymbol{\xi}^T$;

(iii) both $\boldsymbol{f}_T + \boldsymbol{\kappa}_T$ and $\boldsymbol{\kappa}_T$ are continuous convex functions on $\operatorname{co} Y_{\cap}^T \times \operatorname{co} Y_{\cup}^T$.

The adjusted recourse function Ψ_T is given by the optimal value of the maximization problem (5.10) below.

$$\begin{aligned} \sup_{\boldsymbol{x}_T \in \mathbb{R}^{n_T}} \quad & \rho_T(\boldsymbol{x}^T, \boldsymbol{\eta}^T) + \alpha_T(\boldsymbol{\xi}^T) \\ \text{s.t.} \quad & \boldsymbol{f}_T(\boldsymbol{x}^T, \boldsymbol{\xi}^T) \leq \mathbf{0} \end{aligned} \tag{5.10}$$

Notice that the parametric optimization problem (5.10) has the same structure as problem (E.1) in appendix E. Thus, proposition E.5 is applicable implying that there exists a bounding vector $\boldsymbol{D}_T^* \in \mathbb{R}^{r_T}$ for the Lagrange multipliers associated with the explicit constraints in (5.10) uniformly on some neighborhood of Z^T. In particular, by the propositions 5.9 and E.5, there are open (not necessarily convex) sets

$$Z_{\cap}^T \subset \mathbb{R}^{n^{T-1}} \times \mathbb{R}^{L^T} \quad \text{and} \quad Z_{\cup}^T \subset \mathbb{R}^{K^T}$$

such that

(iv) $Z_{\cap}^T \times Z_{\cup}^T$ is a neighborhood of Z^T;

(v) the graph of X_T over $Z_{\cap}^T \times Z_{\cup}^T$ is a subset of $Y_{\cap}^T \times Y_{\cup}^T$;

(vi) the multifunction X_T is non-empty-compact-valued on $Z_{\cap}^T \times Z_{\cup}^T$;

(vii) there exists a bounding vector $\boldsymbol{D}_T^* \in \mathbb{R}^{r_T}$ for the Lagrange multipliers associated with the constraints in (5.12) uniformly over $Z_{\cap}^T \times Z_{\cup}^T$.

Assertion (vii) proves claim (a) for stage T.[4] The existence of a uniform bounding vector for the Lagrange multipliers allows us to invoke the results of appendix C to derive a penalty-based formulation of problem (5.10). By corollary C.2 and the definition of the correction terms (5.6) we may conclude that the optimal value of

$$\sup_{\boldsymbol{x}_T \in \mathbb{R}^{n_T}} \rho_T(\boldsymbol{x}^T, \boldsymbol{\eta}^T) - \langle \boldsymbol{D}_T^*, [\boldsymbol{f}_T(\boldsymbol{x}^T, \boldsymbol{\xi}^T)]^+ + \boldsymbol{\kappa}_T(\boldsymbol{\xi}^T)\rangle \tag{5.11}$$

is given by Ψ_T on $Z_\cap^T \times Z_\cup^T$. Next, let us define an auxiliary objective function

$$\bar{\rho}_T := \begin{cases} \rho_T - \langle \boldsymbol{D}_T^*, [\boldsymbol{f}_T]^+ + \boldsymbol{\kappa}_T\rangle & \text{on} \quad \operatorname{co} Y_\cap^T \times \operatorname{co} Y_\cup^T, \\ +\infty & \text{on} \quad \operatorname{co} Y_\cap^T \times (\operatorname{co} Y_\cup^T)^{\mathrm{c}}, \\ -\infty & \text{everywhere else.} \end{cases}$$

As will be shown below, $\bar{\rho}_T$ is a continuous saddle function on the open convex set $\operatorname{co} Y_\cap^T \times \operatorname{co} Y_\cup^T$. In fact, the profit function ρ_T is a continuous saddle function as implied by assertion (ii). In addition, the penalty term

$$-\langle \boldsymbol{D}_T^*, [\boldsymbol{f}_T]^+ + \boldsymbol{\kappa}_T\rangle = -\langle \boldsymbol{D}_T^*, \max\{\boldsymbol{f}_T + \boldsymbol{\kappa}_T, \boldsymbol{\kappa}_T\}\rangle$$

is continuous, concave, and independent of $\boldsymbol{\eta}^T$ by assertion (iii). Recall that the operator 'max' applied to a set of vectors stands for componentwise maximization. Thus, the auxiliary objective function $\bar{\rho}_T$ is continuous on $\operatorname{co} Y_\cap^T \times \operatorname{co} Y_\cup^T$. Moreover, by construction, $\bar{\rho}_T$ is an extended-real-valued saddle function on the entire underlying space. For the further argumentation we need the sup-projection defined through

$$\bar{\Psi}_T(\boldsymbol{x}^{T-1}, \boldsymbol{\eta}^T, \boldsymbol{\xi}^T) := \sup_{\boldsymbol{x}_T \in \mathbb{R}^{n_T}} \bar{\rho}_T(\boldsymbol{x}^T, \boldsymbol{\eta}^T, \boldsymbol{\xi}^T).$$

The results on sup-projections in Sect. 3.4 guarantee that $\bar{\Psi}_T$ is a saddle function on the entire space (cf. also the related argument in the proof of theorem 3.14). Furthermore, $\bar{\Psi}_T$ is pointwise finite on the open set $Z_\cap^T \times Z_\cup^T$ due to the statements (v) and (vi). The saddle property then implies pointwise finiteness on the convex hull of $Z_\cap^T \times Z_\cup^T$. Pointwise finiteness of a saddle function on an open set is equivalent to continuity and subdifferentiability as implied by [89, theorem 10.1 and theorem 23.4]. Finally, we may conclude that $\Psi_T \equiv \bar{\Psi}_T$ on $Z_\cap^T \times Z_\cup^T$ due to assertion (v) and the construction of $\bar{\Psi}_T$. Consequently, claim (b) is established for stage T, and the basis step is complete.

Assume now that the assertions (a) and (b) have been proved for stage $t+1$. In addition, assume that there exists a function $\bar{\Psi}_{t+1}$ with the postulated properties. Hence, by the assumptions (D2) and (D3) and the induction hypothesis there are open (not necessarily convex) sets

[4] Note that the correction terms have no influence on the Lagrange multipliers since they are independent of the decision variables.

$$Y^t_{\cap} \subset \mathbb{R}^{n^t} \times \mathbb{R}^{L^t} \quad \text{and} \quad Y^t_{\cup} \subset \mathbb{R}^{K^t}$$

such that

(i)* $Y^t_{\cap} \times Y^t_{\cup}$ is a neighborhood of Y^t;

(ii)* both ρ_t and $(E_t \bar{\Psi}_{t+1})$ are continuous saddle functions on $\operatorname{co} Y^t_{\cap} \times \operatorname{co} Y^t_{\cup}$ being concave in $\boldsymbol{x}^t$, convex in $\boldsymbol{\eta}^t$, and constant in $\boldsymbol{\xi}^t$;

(iii)* both $\boldsymbol{f}_t + \boldsymbol{\kappa}_t$ and $\boldsymbol{\kappa}_t$ are continuous convex functions on $\operatorname{co} Y^t_{\cap} \times \operatorname{co} Y^t_{\cup}$;

(iv)* $(E_t \Psi_{t+1}) \equiv (E_t \bar{\Psi}_{t+1})$ on $Y^t_{\cap} \times Y^t_{\cup}$.

Due to assertion (iv)* the optimal value of the maximization problem (5.12) below coincides with the adjusted recourse function Ψ_t at least on Z^t.

$$\sup_{\boldsymbol{x}_t \in \mathbb{R}^{n_t}} \rho_t(\boldsymbol{x}^t, \boldsymbol{\eta}^t) + \alpha_t(\boldsymbol{\xi}^t) + (E_t \bar{\Psi}_{t+1})(\boldsymbol{x}^t, \boldsymbol{\eta}^t, \boldsymbol{\xi}^t) \qquad (5.12)$$
$$\text{s.t.} \quad \boldsymbol{f}_t(\boldsymbol{x}^t, \boldsymbol{\xi}^t) \leq \boldsymbol{0}$$

As in the basis step, proposition E.5 is applicable implying that there exists a bounding vector $\boldsymbol{D}^*_t \in \mathbb{R}^{r_t}$ for the Lagrange multipliers associated with the explicit constraints in (5.12) uniformly over a neighborhood of Z^t. In particular, by the propositions 5.9 and E.5 there are open (not necessarily convex) sets

$$Z^t_{\cap} \subset \mathbb{R}^{n^{t-1}} \times \mathbb{R}^{L^t} \quad \text{and} \quad Z^t_{\cup} \subset \mathbb{R}^{K^t}$$

such that

(v)* $Z^t_{\cap} \times Z^t_{\cup}$ is a neighborhood of Z^t;

(vi)* the graph of X_t over $Z^t_{\cap} \times Z^t_{\cup}$ is a subset of $Y^t_{\cap} \times Y^t_{\cup}$;

(vii)* the multifunction X_t is non-empty-compact-valued on $Z^t_{\cap} \times Z^t_{\cup}$;

(viii)* there exists a bounding vector $\boldsymbol{D}^*_t \in \mathbb{R}^{r_t}$ for the Lagrange multipliers associated with the constraints in (5.12) uniformly over $Z^t_{\cap} \times Z^t_{\cup}$.

The assertions (iv)* and (viii)* prove claim (a) for stage t.[5] Moreover, the assertions (iv)* and (vi)* imply that the optimal value of (5.12) is given by Ψ_t for all parameters in $Z^t_{\cap} \times Z^t_{\cup}$. The existence of a uniform bounding vector for the Lagrange multipliers allows us to derive a penalty-based formulation of problem (5.12). By corollary C.2 and the definition of the correction terms (5.6), we may conclude that the optimal value of

[5]Use proposition E.6 in the appendix, and observe that the correction terms have no influence on the Lagrange multipliers.

$$\sup_{\boldsymbol{x}_t \in \mathbb{R}^{n_t}} \rho_t(\boldsymbol{x}^t, \boldsymbol{\eta}^t) + (E_t \bar{\Psi}_{t+1})(\boldsymbol{x}^t, \boldsymbol{\eta}^t, \boldsymbol{\xi}^t) - \langle \boldsymbol{D}_t^*, [\boldsymbol{f}_t(\boldsymbol{x}^t, \boldsymbol{\xi}^t)]^+ + \boldsymbol{\kappa}_t(\boldsymbol{\xi}^t) \rangle \quad (5.13)$$

is given by Ψ_t on $Z_\cap^t \times Z_\cup^t$. Next, let us define an auxiliary objective function

$$\bar{\rho}_t := \begin{cases} \rho_t + (E_t \bar{\Psi}_{t+1}) - \langle \boldsymbol{D}_t^*, [\boldsymbol{f}_t]^+ + \boldsymbol{\kappa}_t \rangle & \text{on} \quad \operatorname{co} Y_\cap^t \times \operatorname{co} Y_\cup^t, \\ +\infty & \text{on} \quad \operatorname{co} Y_\cap^t \times (\operatorname{co} Y_\cup^t)^c, \\ -\infty & \text{everywhere else.} \end{cases}$$

By the assertions (ii)* and (iii)* and by the induction hypothesis, $\bar{\rho}_t$ represents an extended-real-valued saddle function on the entire underlying space being continuous on the open convex set $\operatorname{co} Y_\cap^t \times \operatorname{co} Y_\cup^t$. For the further argumentation we need the sup-projection

$$\bar{\Psi}_t(\boldsymbol{x}^{t-1}, \boldsymbol{\eta}^t, \boldsymbol{\xi}^t) := \sup_{\boldsymbol{x}_t \in \mathbb{R}^{n_t}} \bar{\rho}_t(\boldsymbol{x}^t, \boldsymbol{\eta}^t, \boldsymbol{\xi}^t).$$

By the same reasoning as in the basis step, one shows $\bar{\Psi}_t$ to be a saddle function on the entire space which is continuous on the convex hull of $Z_\cap^t \times Z_\cup^t$. Finally, we may conclude that $\Psi_t \equiv \bar{\Psi}_t$ on $Z_\cap^t \times Z_\cup^t$ due to assertion (vi)* and the construction of $\bar{\Psi}_t$. Consequently, claim (b) is established for stage t, and the induction step is complete. □

In the proof of proposition 5.12 we show existence of the upper bounding vectors $\{\boldsymbol{D}_t^*\}_{t \in \tau}$ by means of a backward recursion scheme. In practical applications, the bounding vectors of the Lagrange multipliers may be hard to find. Suitable estimates frequently follow from a good understanding of the underlying models (see also Chap. 6).

Proposition 5.13. *Consider a multistage stochastic program subject to the conditions (D1)–(D5), and define the adjusted recourse problem as in (5.9). Then, we may conclude:*

(a) $\Psi_t^l \leq \Psi_t \leq \Psi_t^u$ *on* Z^t $\forall t \in \tau$;

(b) $(E^l \Psi_0^l) \leq (E \Psi_0) \leq (E^u \Psi_0^u)$;

(c) $\Psi_t = \Phi_t + A_t$ *and* $\Psi_t^d = \Phi_t^d + A_t^d$ $\forall d \in \{l, u\}$, $t \in \tau$;

(d) $(E \Psi_0) = (E \Phi_0) + (E A_0)$ *and* $(E^d \Psi_0^d) = (E^d \Phi_0^d) + (E^d A_0^d)$ $\forall d \in \{l, u\}$.

Proof. By proposition 5.12 the recourse function Ψ_t of the adjusted recourse problem (5.9) is a subdifferentiable saddle function on a neighborhood of its natural domain Z^t for each $t \in \tau$. Hence, theorem 4.7 applies to the adjusted recourse problem (5.9) without modification of the proof. This implies that Ψ_t is squeezed between the auxiliary value functions Ψ_t^l and Ψ_t^u for all $t \in \tau$, and thus the assertions (a) and (b) are established. The statements (c) and (d) are straightforward and follow directly from the definition of the adjusted recourse problem and the fact that the correction terms are independent of the decision variables. □

Theorem 5.14 (Adjusted Bounds II). *Consider a stochastic program satisfying the regularity conditions (D1)–(D5). Then, the correction terms* $\{\alpha_t\}_{t\in\tau}$ *specified in (5.6) can be used to construct (finite) bounds on the recourse function:*

$$\Phi_t^l + A_t^l - A_t \leq \Phi_t \leq \Phi_t^u + A_t^u - A_t \quad \text{on } Z^t \text{ for all } t \in \tau.$$

Analogously, the optimal objective function value has bounds:

$$(E^l \Phi_0^l) + (E^l A_0^l) - (EA_0) \leq (E\Phi_0) \leq (E^u \Phi_0^u) + (E^u A_0^u) - (EA_0).$$

Proof. The claim follows directly from proposition 5.13. □

Theorem 5.15 (Convergence). *Consider a stochastic program satisfying the regularity conditions (D1)–(D5). If the barycentric probability measures are regularly refined in the sense of (4.37), then the random variables* A_t^l *and* A_t^u *converge to* A_t *uniformly on* Ξ^t *while the auxiliary recourse functions* Φ_t^l *and* Φ_t^u *converge to* Φ_t *uniformly on* Z^t *for all* $t \in \tau$.

Proof. By assumption, the correction term α_t is uniformly continuous on Ξ^t for every $t \in \tau$. Thus, convergence of A_t^l and A_t^u to A_t uniformly on Ξ^t $(t \in \tau)$ follows from the weak convergence of the discrete barycentric measures to the original probability measure. By proposition 5.12 the recourse function Ψ_t of the adjusted recourse problem (5.9) is a continuous saddle function on its natural domain Z^t for $t \in \tau$. Thus, one can easily verify that theorem 4.8 applies to the adjusted recourse problem (5.9) without modification of the proof. This observation implies convergence of Ψ_t^l and Ψ_t^u to Ψ_t uniformly on Z^t for all $t \in \tau$. Combining the above results, by proposition 5.13 (c) we have convergence of Φ_t^l and Φ_t^u to Φ_t uniformly on Z^t for all $t \in \tau$. □

5.3 Synthesis of Results

In this section we combine the previous results to extend the scope of the barycentric approximation scheme to an even broader class of convex optimization problems. Concretely speaking, we study stochastic programs of the form (3.1) with regularizable profit functions in the sense of definition 5.1 and regularizable constraint functions in the sense of definition 5.7. Such problems are supposed to satisfy the following regularity conditions:

(E1) = (B1); (E2) = (C2); (E3) = (D3); (E4) = (B4); (E5) = (B5).

Here, neither the profit nor the constraint functions are assumed to be convex in the stochastic parameters. For every $t \in \tau$, α_t^o denotes a suitable correction term as in definition 5.1, which is assumed to be given. Thus, by condition (E2), the regularized profit functions $\{\rho_t + \alpha_t^{\mathrm{o}}\}_{t\in\tau}$ satisfy the more restrictive condition (B2). Consequently, the stochastic program

$$\begin{aligned} \sup_{\boldsymbol{x}\in\mathcal{N}_n} \int_{\Theta\times\Xi} & \left[\sum_{t=0}^{T} \rho_t(\boldsymbol{x}^t,\boldsymbol{\eta}^t) + \alpha_t^{\mathrm{o}}(\boldsymbol{\eta}^t)\right] dP(\boldsymbol{\eta},\boldsymbol{\xi}) \\ \text{s.t.} \quad & \boldsymbol{f}_t(\boldsymbol{x}^t,\boldsymbol{\xi}^t) \le \mathbf{0} \quad P\text{-a.s.} \quad t\in\tau \end{aligned} \tag{5.14}$$

is subject to the regularity conditions (D1)–(D5). This observation allows us to apply the specific results of Sect. 5.2. Above all, we may conclude that there are correction terms α_t^{r} as in (5.6); notice that the involved bounding vectors for the Lagrange multipliers exist due to proposition 5.12 (a). Then, for all $t\in\tau$ define a combined *correction term* as

$$\alpha_t(\boldsymbol{\eta}^t,\boldsymbol{\xi}^t) := \alpha_t^{\mathrm{o}}(\boldsymbol{\eta}^t) + \alpha_t^{\mathrm{r}}(\boldsymbol{\xi}^t). \tag{5.15}$$

Obviously, α_t is a continuous saddle function on a neighborhood of $\Theta^t\times\Xi^t$ being convex in $\boldsymbol{\eta}^t$ and concave in $\boldsymbol{\xi}^t$. As usual, let us introduce three sequences of additional random variables. For $t\in\tau$ define

$$\begin{aligned} A_t(\boldsymbol{\eta}^t,\boldsymbol{\xi}^t) &:= E_t\left[\sum_{s=t}^{T}\alpha_s(\boldsymbol{\eta}^s,\boldsymbol{\xi}^s)\right], \\ A_t^l(\boldsymbol{\eta}^t,\boldsymbol{\xi}^t) &:= E_t^l\left[\sum_{s=t}^{T}\alpha_s(\boldsymbol{\eta}^s,\boldsymbol{\xi}^s)\right], \\ A_t^u(\boldsymbol{\eta}^t,\boldsymbol{\xi}^t) &:= E_t^u\left[\sum_{s=t}^{T}\alpha_s(\boldsymbol{\eta}^s,\boldsymbol{\xi}^s)\right], \end{aligned} \tag{5.16}$$

and for the sake of transparent notation introduce three constants

$$\begin{aligned} (E\,A_0) &:= E\ \left[A_0(\eta_0,\xi_0)\right], \\ (E^l A_0^l) &:= E^l\left[A_0^l(\eta_0,\xi_0)\right], \\ (E^u A_0^u) &:= E^u\left[A_0^u(\eta_0,\xi_0)\right]. \end{aligned} \tag{5.17}$$

The random variables (5.16) as well as the constants (5.17) will be referred to as *conditional correction terms* below. It is important to notice that the conditional correction terms are computationally accessible. Evaluation of A_t^l and A_t^u requires calculation of a finite sum, while A_t can be evaluated by means of numerical integration techniques. Since α_t is continuous for every $t\in\tau$, A_t, A_t^l, and A_t^u are continuous and bounded on $\Theta^t\times\Xi^t$. Moreover, by the saddle structure of the correction terms, we find $A_t^l\le A_t\le A_t^u$ on $\Theta^t\times\Xi^t$, although A_t^l and A_t^u do not necessarily exhibit a saddle shape (use a similar argument as in the proof of theorem 4.7). Following (5.9), we associate to (5.14) an *adjusted recourse problem* with profit functions

$$\rho_t + \alpha_t^{\mathrm{o}} + \alpha_t^{\mathrm{r}} \equiv \rho_t + \alpha_t.$$

Since the correction terms are independent of the decision variables, it is obvious that for each $t \in \tau$ the optimal value function of the adjusted recourse problem reduces to $\Phi_t + A_t$ (cf. also the assertions (c) and (d) of proposition 5.13). Furthermore, arguing as in proposition 5.13 (a) and (b), one can show that theorem 4.7 applies to the adjusted recourse problem associated with (5.14). Thus, the auxiliary value functions $\Phi_t^l + A_t^l$ and $\Phi_t^u + A_t^u$ obtained via barycentric discretization provide upper and lower bounds on the original value function of the adjusted recourse problem. The above results culminate in the following theorem, which is clearly a generalization of the theorems 5.5 and 5.14.

Theorem 5.16 (Adjusted Bounds III). *Consider a stochastic program satisfying the regularity conditions (E1)–(E5), and let $\{\alpha_t\}_{t\in\tau}$ be suitable correction terms as in (5.15). Then, we find*

$$\Phi_t^l + A_t^l - A_t \leq \Phi_t \leq \Phi_t^u + A_t^u - A_t \quad \text{on } Z^t \text{ for all } t \in \tau.$$

Analogously, the optimal objective function value has bounds:

$$(E^l \Phi_0^l) + (E^l A_0^l) - (EA_0) \leq (E\Phi_0) \leq (E^u \Phi_0^u) + (E^u A_0^u) - (EA_0).$$

As in the previous sections, the upper and lower bounds on the recourse functions become tighter as the barycentric measures are suitably refined. Thus, one can easily prove the following convergence result.

Theorem 5.17 (Convergence). *Consider a stochastic program satisfying the regularity conditions (E1)–(E5). If the barycentric probability measures are regularly refined in the sense of (4.37), then the random variables A_t^l and A_t^u converge to A_t uniformly on $\Theta^t \times \Xi^t$ while the auxiliary recourse functions Φ_t^l and Φ_t^u converge to Φ_t uniformly on Z^t for all $t \in \tau$.*

5.4 Linear Stochastic Programs

Linear stochastic programs were first studied by Danzig [17, 18]. One of the reasons for their wide use is the existence of powerful solution algorithms, which exploit specific structural properties. Any *linear* multistage stochastic program can be brought to the form

$$\begin{aligned} &\sup_{\boldsymbol{x}\in\mathcal{N}_n} \int_{\Theta\times\Xi} \left[\sum_{t=0}^{T} \langle \boldsymbol{c}_t^*(\boldsymbol{\eta}^t), \boldsymbol{x}_t \rangle \right] dP(\boldsymbol{\eta}, \boldsymbol{\xi}) && (5.18) \\ &\text{s.t.} \quad W_t(\boldsymbol{\xi}^t)\boldsymbol{x}_t + T_t(\boldsymbol{\xi}^t)\boldsymbol{x}_{t-1} \sim_t \boldsymbol{h}_t(\boldsymbol{\xi}^t) \quad P\text{-a.s.} \quad t \in \tau. \end{aligned}$$

In contrast to general convex stochastic programs, the profit and constraint functions are now linear in the decision variables. However, we do not require linearity in the stochastic parameters. When dealing with linear stochastic programs, it proves useful to work explicitly with inequality (less-or-equal, greater-or-equal) and equality constraints. Thus, for notational convenience we introduce in every stage an r_t^{eff}-dimensional 'vector' $\sim_t$ of binary relations, each of whose entries is either '$\leq$' or '$\geq$' for inequalities, or '$=$' for equalities. Of course, the stochastic program (5.18) can be brought to the form (3.1) if the equality constraints are replaced by two opposing inequality constraints, and the 'greater-or-equal' constraints are multiplied by -1. Notice that the constraints in (5.18) only couple neighboring decision stages. This can always be enforced, i.e. dependencies across more than one decision stage can systematically be eliminated by introducing additional decision variables. For every $t \in \tau$, the $r_t^{\mathrm{eff}} \times n_t$ matrix W_t is termed *recourse matrix* and may generally depend on the stochastic parameters $\boldsymbol{\xi}^t$. Similarly, the $r_t^{\mathrm{eff}} \times n_{t-1}$ matrix T_t is referred to as *technology matrix* in literature. Obviously, the technology matrix determines the interperiodical coupling and generally represents a random object, as it may depend on $\boldsymbol{\xi}^t$. Moreover, the *right hand side (rhs) vector* $\boldsymbol{h}_t$ and the vector of *objective function coefficients* $\boldsymbol{c}_t^*$ are random too, since they depend on $\boldsymbol{\xi}^t$ and $\boldsymbol{\eta}^t$, respectively.

After a possible reordering, the constraints in (5.18) can be written as

$$\begin{array}{rcl} W_t^{\mathrm{le}}(\boldsymbol{\xi}^t)\,\boldsymbol{x}_t + T_t^{\mathrm{le}}(\boldsymbol{\xi}^t)\,\boldsymbol{x}_{t-1} - \boldsymbol{h}_t^{\mathrm{le}}(\boldsymbol{\xi}^t) & \leq & \mathbf{0}, \\ W_t^{\mathrm{ge}}(\boldsymbol{\xi}^t)\,\boldsymbol{x}_t + T_t^{\mathrm{ge}}(\boldsymbol{\xi}^t)\,\boldsymbol{x}_{t-1} - \boldsymbol{h}_t^{\mathrm{ge}}(\boldsymbol{\xi}^t) & \geq & \mathbf{0}, \\ W_t^{\mathrm{eq}}(\boldsymbol{\xi}^t)\,\boldsymbol{x}_t + T_t^{\mathrm{eq}}(\boldsymbol{\xi}^t)\,\boldsymbol{x}_{t-1} - \boldsymbol{h}_t^{\mathrm{eq}}(\boldsymbol{\xi}^t) & = & \mathbf{0}, \end{array}$$

and for any fixed $\boldsymbol{\xi}^t \in \mathbb{R}^{L^t}$ we have

$$\begin{array}{l} W_t^{\mathrm{le}} \in \mathbb{R}^{r_t^{\mathrm{le}} \times n_t},\; T_t^{\mathrm{le}} \in \mathbb{R}^{r_t^{\mathrm{le}} \times n_{t-1}},\; \boldsymbol{h}_t^{\mathrm{le}} \in \mathbb{R}^{r_t^{\mathrm{le}}}, \\ W_t^{\mathrm{ge}} \in \mathbb{R}^{r_t^{\mathrm{ge}} \times n_t},\; T_t^{\mathrm{ge}} \in \mathbb{R}^{r_t^{\mathrm{ge}} \times n_{t-1}},\; \boldsymbol{h}_t^{\mathrm{ge}} \in \mathbb{R}^{r_t^{\mathrm{ge}}}, \\ W_t^{\mathrm{eq}} \in \mathbb{R}^{r_t^{\mathrm{eq}} \times n_t},\; T_t^{\mathrm{eq}} \in \mathbb{R}^{r_t^{\mathrm{eq}} \times n_{t-1}},\; \boldsymbol{h}_t^{\mathrm{eq}} \in \mathbb{R}^{r_t^{\mathrm{eq}}}. \end{array}$$

The dimensions should match up properly, i.e. $r_t^{\mathrm{eff}} = r_t^{\mathrm{le}} + r_t^{\mathrm{ge}} + r_t^{\mathrm{eq}}$. In this case, the vector of binary relations reads

$$\sim_t = (\underbrace{\leq, \ldots, \leq}_{r_t^{\mathrm{le}}}, \underbrace{\geq, \ldots, \geq}_{r_t^{\mathrm{ge}}}, \underbrace{=, \ldots, =}_{r_t^{\mathrm{eq}}}).$$

With the above conventions it is easy to see that (5.18) is equivalent to a stochastic program of the form (3.1) with profit functions

$$\rho_t(\boldsymbol{x}^t, \boldsymbol{\eta}^t) := \langle \boldsymbol{c}_t^*(\boldsymbol{\eta}^t), \boldsymbol{x}_t \rangle \tag{5.19}$$

and r_t-dimensional constraint functions

$$f_t(\boldsymbol{x}^t, \boldsymbol{\xi}^t) := \begin{pmatrix} W_t^{\text{le}}(\boldsymbol{\xi}^t)\,\boldsymbol{x}_t + T_t^{\text{le}}(\boldsymbol{\xi}^t)\,\boldsymbol{x}_{t-1} - \boldsymbol{h}_t^{\text{le}}(\boldsymbol{\xi}^t) \\ -W_t^{\text{ge}}(\boldsymbol{\xi}^t)\,\boldsymbol{x}_t - T_t^{\text{ge}}(\boldsymbol{\xi}^t)\,\boldsymbol{x}_{t-1} + \boldsymbol{h}_t^{\text{ge}}(\boldsymbol{\xi}^t) \\ W_t^{\text{eq}}(\boldsymbol{\xi}^t)\,\boldsymbol{x}_t + T_t^{\text{eq}}(\boldsymbol{\xi}^t)\,\boldsymbol{x}_{t-1} - \boldsymbol{h}_t^{\text{eq}}(\boldsymbol{\xi}^t) \\ -W_t^{\text{eq}}(\boldsymbol{\xi}^t)\,\boldsymbol{x}_t - T_t^{\text{eq}}(\boldsymbol{\xi}^t)\,\boldsymbol{x}_{t-1} + \boldsymbol{h}_t^{\text{eq}}(\boldsymbol{\xi}^t) \end{pmatrix}, \tag{5.20}$$

where $r_t = r_t^{\text{le}} + r_t^{\text{ge}} + 2\,r_t^{\text{eq}}$ for all $t \in \tau$. The latter optimization problem is called the *normal form* of the linear stochastic program (5.18). As will become clear below, the formulation (5.18) accentuates the intrinsic primal-dual symmetry properties of linear programs, whereas the normal form is needed for comparison with our previous results. In the remainder of this section the regularity conditions (B1), (B4), and (B5) are assumed to hold. Now, let us briefly discuss under which conditions on $\boldsymbol{c}_t^*$, W_t, T_t, and $\boldsymbol{h}_t$ the normal form of the linear stochastic program (5.18) can be expected to comply with (B2) and (B3).

Proposition 5.18. *Consider a linear stochastic program which complies with the regularity conditions (B1), (B4), and (B5). Then, the corresponding normal form satisfies condition (B2) if and only if the vector-valued functions $\boldsymbol{c}_t^*$ are continuous, and there is a convex neighborhood of Θ^t where the following implications hold ($t \in \tau$, $i = 1, \ldots, n_t$):*

$$\begin{aligned} &(i)\ \forall \boldsymbol{x}_t \in X_t(Z^t): & x_{t,i} > 0 \ &\Rightarrow\ c_{t,i}^* \text{ is convex in } \boldsymbol{\eta}^t, \\ &(ii)\ \forall \boldsymbol{x}_t \in X_t(Z^t): & x_{t,i} < 0 \ &\Rightarrow\ c_{t,i}^* \text{ is concave in } \boldsymbol{\eta}^t, \\ &(iii)\ \exists \boldsymbol{x}_t \in X_t(Z^t): & x_{t,i} = 0 \ &\Rightarrow\ c_{t,i}^* \text{ is linear affine in } \boldsymbol{\eta}^t. \end{aligned} \tag{5.21}$$

Proof. The proof is merely a careful check of definitions. Recall that $X_t(Z^t)$ is the image of the feasible set mapping X_t over Z^t, which equals the projection of Y^t on $\mathbb{R}^{n_t}$. □

Proposition 5.19. *Consider a linear stochastic program which complies with the regularity conditions (B1), (B4), and (B5). Then, the corresponding normal form satisfies condition (B3) if and only if the vector- and matrix-valued mappings $\boldsymbol{h}_t$, W_t, and T_t are continuous, and there is a convex neighborhood of Ξ^t where we have ($t \in \tau$, $i = 1, \ldots, r_t^{\text{eff}}$):*

$$\begin{aligned} &(i)\ \text{the } i\text{'th entry of } \sim_t \text{ is `}\geq\text{'} \ \Rightarrow\ h_{t,i} \text{ is convex in } \boldsymbol{\xi}^t, \\ &(ii)\ \text{the } i\text{'th entry of } \sim_t \text{ is `}\leq\text{'} \ \Rightarrow\ h_{t,i} \text{ is concave in } \boldsymbol{\xi}^t, \\ &(iii)\ \text{the } i\text{'th entry of } \sim_t \text{ is `}=\text{'} \ \Rightarrow\ h_{t,i} \text{ is linear affine in } \boldsymbol{\xi}^t, \end{aligned} \tag{5.22}$$

(iv) the matrices W_t and T_t are deterministic,

(v) the cone $\{\boldsymbol{x}_t | W_t \boldsymbol{x}_t \sim_t \boldsymbol{0}\}$ contains only the zero vector.

Proof. First, we show sufficiency. Independence of the recourse matrix W_t and technology matrix T_t from the stochastic parameters follows immediately from proposition 5.8. The properties (i), (ii), and (iii) of the rhs vectors $\boldsymbol{h}_t$ are then straightforward. Moreover, property (v) follows from boundedness of the feasible set mapping X_t on a neighborhood of Z^t.

Let us next prove necessity. Under the given assumptions, the constraint function (5.20) is obviously continuous and convex in the decision variables. Moreover, it is jointly convex in $(\boldsymbol{x}^t, \boldsymbol{\xi}^t)$ on a convex neighborhood of Y^t. It remains to be shown that the associated feasible set mapping X_t is bounded on a neighborhood of Z^t for all $t \in \tau$. As a byproduct, we will prove compactness of the generalized feasible sets.

We prove compactness of Y^t and Z^t by induction with respect to t. The basis step is trivial since Z^0 is compact by assumption (B1). Next, assume Z^t to be compact for some $t \in \tau$. By assumption, the recession cone of the feasible set $X_t(\boldsymbol{x}^{t-1}, \boldsymbol{\xi}^t)$ is given by the singleton $\{\boldsymbol{0}\}$ for every outcome and decision history in a neighborhood of Z^t. Thus, the feasible set mapping X_t is pointwise bounded on a neighborhood of Z^t. Uniform boundedness follows from continuity of the rhs vector $\boldsymbol{h}_t$, independence of W_t and T_t from the stochastic parameters, and compactness of Z^t. Next, recall that Y^t is the graph of the bounded multifunction X_t over Z^t. Therefore, Y^t is bounded. In addition, continuity of the constraint functions guarantees compactness of Y^t. If $t = T$, we are finished. Otherwise, compactness of Y^t implies compactness of $Z^{t+1} = Y^t \times \Theta_{t+1} \times \Xi_{t+1}$. This notion completes the induction step, and thus the claim follows. □

Below we carry over the results of Sect. 5.3 to the case of linear stochastic programs with nonconvex objective function coefficients and rhs vectors. Thereby, we are naturally led to the study of functions which are representable as a difference of two convex functions.

5.4.1 D.c. Functions

Let Ω be a convex subset of a finite-dimensional Euclidean space, and denote by $CO(\Omega; \mathbb{R}^r)$ the cone of componentwise convex functions from Ω into $\mathbb{R}^r$.

Definition 5.20. *A function $\boldsymbol{f} : \Omega \to \mathbb{R}^r$ is called d.c. (abbreviation for difference of convex functions) if there are two functions $\boldsymbol{\kappa}^+, \boldsymbol{\kappa}^- \in CO(\Omega; \mathbb{R}^r)$ such that*

$$\boldsymbol{f}(\omega) = \boldsymbol{\kappa}^+(\omega) - \boldsymbol{\kappa}^-(\omega) \quad \forall \omega \in \Omega.$$

Notice that the decomposition of a d.c. function is never unique. In fact, by adding the same convex mapping to both $\boldsymbol{\kappa}^+$ and $\boldsymbol{\kappa}^-$, one easily obtains a new decomposition which fits the above definition. Recently, d.c. functions have experienced a great deal of attention in the field of global optimization [40,55,80]. A survey of the properties of real-valued d.c. functions can be found in [53]. Here, we recall only a few properties of vector-valued d.c. functions, which are important in the present context. The class of d.c. functions $\Omega \to \mathbb{R}^r$, denoted by $DC(\Omega;\mathbb{R}^r)$, is clearly the vector space generated by the cone of convex functions from Ω into $\mathbb{R}^r$.

$$DC(\Omega;\mathbb{R}^r) = CO(\Omega;\mathbb{R}^r) - CO(\Omega;\mathbb{R}^r)$$

Some properties of $\boldsymbol{f} \in DC(\Omega;\mathbb{R}^r)$ are directly inherited from those of convex functions. For instance, $\boldsymbol{f}$ is locally Lipschitz on the interior of Ω, and the derivative of $\boldsymbol{f}$ does exist almost everywhere with respect to the Lebesgue measure on Ω. D.c. functions $\boldsymbol{f} : \mathbb{R} \to \mathbb{R}^r$ of one real variable have a simple internal characterization: $\boldsymbol{f}$ is d.c. if and only if it is locally Lipschitz and its derivative $\boldsymbol{f}'$ (defined almost everywhere) is of finite variation. In other words, d.c. functions on $\mathbb{R}$ are precisely indefinite integrals of functions with locally bounded variation [53]. The following example of a Lipschitz function which is not d.c. originates from Shapiro [100].

$$f : \begin{cases} \mathbb{R} \to \mathbb{R} \\ \omega \mapsto \min_{n \in \mathbb{N}} |\omega - \frac{1}{n}| \end{cases}$$

Obviously, f is Lipschitz with modulus 1, and the derivative f' exists almost everywhere. However, f' oscillates between $+1$ and -1 infinitely often in any neighborhood of the origin. Thus, f' is not of bounded variation implying that f is not d.c. It is worthwhile to remark that no simple and useful internal characterization of d.c. functions of more than just one real variable is known.

The class of twice continuously differentiable functions from Ω into $\mathbb{R}^r$, i.e. $C^2(\Omega;\mathbb{R}^r)$, is a linear subspace of $DC(\Omega;\mathbb{R}^r)$. This can easily be proved if Ω is compact (cf. also the related propositions 5.3 and 5.11, whose proofs rely on similar arguments). However, the statement remains true for Ω open or unbounded, as pointed out by Hartman [52]. Moreover, let $C^0(\Omega;\mathbb{R}^r)$ be the space of continuous mappings $\Omega \to \mathbb{R}^r$, endowed with the topology of uniform convergence with respect to the Euclidean norm in $\mathbb{R}^r$. Then, $DC(\Omega;\mathbb{R}^r)$ is dense in $C^0(\Omega;\mathbb{R}^r)$ given that Ω is compact. This follows directly from the Stone-Weierstrass theorem and the fact that $DC(\Omega;\mathbb{R}^r)$ contains all vector-valued polynomials.

5.4.2 Generalized Bounds for Linear Stochastic Programs

In the remainder of this section we study linear multistage stochastic programs of the form (5.18) which satisfy the following regularity conditions:

(F1) = (B1);
(F2) the objective function coefficients c_t^* are globally continuous and d.c. on a convex neighborhood of Θ^t, $t \in \tau$;
(F3) – the rhs vector h_t is globally continuous and d.c. on a convex neighborhood of Ξ^t, $t \in \tau$;
– the matrices W_t and T_t are globally continuous and constant on a convex neighborhood of Ξ^t, $t \in \tau$;
– the recession cone $\{x_t | W_t x_t \sim_t \mathbf{0}\}$ is given by $\{\mathbf{0}\}$ on a convex neighborhood of Ξ^t, $t \in \tau$;
(F4) = (B4); (F5) = (B5).

As the random vector c_t^* is d.c., there are two convex mappings κ_t^{*+} and κ_t^{*-} on a closed convex neighborhood of Θ^t such that

$$c_t^* = \kappa_t^{*+} - \kappa_t^{*-} \quad \forall t \in \tau. \tag{5.23}$$

By the Tietze extension theorem [110, p. 103], both κ_t^{*+} and κ_t^{*-} can be extended to globally continuous functions which satisfy (5.23) on all of $\mathbb{R}^{K^t}$. Similarly, since h_t is d.c., we may conclude that there are two convex mappings κ_t^+ and κ_t^- on a closed convex neighborhood of Ξ^t with

$$h_t = \kappa_t^+ - \kappa_t^- \quad \forall t \in \tau. \tag{5.24}$$

As in the case of the objective function coefficients, κ_t^+ and κ_t^- have continuous extensions which satisfy (5.24) on all of $\mathbb{R}^{L^t}$. In the next proposition, we establish a link to the results of Sect. 5.3.

Proposition 5.21. *The linear stochastic program (5.18) satisfies (F1)–(F5) if and only if the corresponding normal form satisfies (E1)–(E5).*

Proof. First, we show necessity. If the linear stochastic program (5.18) satisfies the regularity conditions (F1)–(F5), then the rhs vectors can be decomposed as in (5.24). Next, rewrite $\kappa_t^\pm$ as $(\kappa_t^{\text{le}\pm}, \kappa_t^{\text{ge}\pm}, \kappa_t^{\text{eq}\pm})$. This representation reflects the classification of constraints by the attributes 'less-or-equal', 'greater-or-equal', and 'equality'. With this convention, one can define a globally continuous mapping

$$\kappa_t := (\kappa_t^{\text{le}+}, \kappa_t^{\text{ge}-}, \kappa_t^{\text{eq}+}, \kappa_t^{\text{eq}-}),$$

which is componentwise convex on a convex neighborhood of Ξ^t. Now, consider the constraint function f_t defined in (5.20). As the involved recourse and technology matrices are deterministic by hypothesis, it can easily be verified that $f_t + \kappa_t$ is jointly convex in (x^t, ξ^t) and constant in η^t on a convex neighborhood of Y^t. Thus, the constraint function (5.20) is regularizable in the sense of definition 5.7. Next, arguing as in the proof of proposition 5.19, one can show that Y^t is compact, and the feasible set mapping X_t is bounded on a neighborhood of Z^t for every $t \in \tau$. This proves (E3). Compactness of Y^t implies that there exist nonnegative bounding vectors $\boldsymbol{X}_t^+$ and $\boldsymbol{X}_t^-$ such that

$$-\boldsymbol{X}_t^- < \boldsymbol{x}_t < \boldsymbol{X}_t^+ \quad \forall \boldsymbol{x}_t \in X_t(Z^t).$$

Recall that $X_t(Z^t)$ is the projection of Y^t on $\mathbb{R}^{n_t}$ and thus compact. By hypothesis, the objective function coefficients can be decomposed as in (5.23). The involved mappings $\boldsymbol{\kappa}_t^{*+}$ and $\boldsymbol{\kappa}_t^{*-}$ are then used to define suitable correction terms α_t^{o} as in definition 5.1.

$$\alpha_t^{\text{o}}(\boldsymbol{\eta}^t) := \langle \boldsymbol{\kappa}_t^{*-}(\boldsymbol{\eta}^t), \boldsymbol{X}_t^+ \rangle + \langle \boldsymbol{\kappa}_t^{*+}(\boldsymbol{\eta}^t), \boldsymbol{X}_t^- \rangle$$

Let ρ_t be the profit function (5.19). Then, we find

$$\rho_t(\boldsymbol{x}^t, \boldsymbol{\eta}^t) + \alpha_t^{\text{o}}(\boldsymbol{\eta}^t) = \langle \boldsymbol{\kappa}_t^{*-}(\boldsymbol{\eta}^t), \boldsymbol{X}_t^+ - \boldsymbol{x}_t \rangle + \langle \boldsymbol{\kappa}_t^{*+}(\boldsymbol{\eta}^t), \boldsymbol{x}_t + \boldsymbol{X}_t^- \rangle.$$

Both terms on the right hand side of the above equation are manifestly linear affine in $\boldsymbol{x}^t$, convex in $\boldsymbol{\eta}^t$, and constant in $\boldsymbol{\xi}^t$ on a convex neighborhood of Y^t. Thus, the profit function (5.19) is regularizable in the sense of definition 5.1, and condition (E2) holds true. In conclusion, the normal form of the linear stochastic program (5.18) satisfies all conditions (E1)–(E5).

Next, we establish sufficiency. By assumption, the normal form of (5.18) is subject to (E1)–(E5). Thus, there is a mapping α_t^{o} as in definition 5.1 such that

$$\langle \boldsymbol{c}_t^*(\boldsymbol{\eta}^t), \boldsymbol{x}_t \rangle + \alpha_t^{\text{o}}(\boldsymbol{\eta}^t) \tag{5.25}$$

is a saddle function on a convex neighborhood of Y^t being linear affine in $\boldsymbol{x}^t$, convex in $\boldsymbol{\eta}^t$, and constant in $\boldsymbol{\xi}^t$. Next, choose an arbitrary $\boldsymbol{y}_t \in X_t(Z^t)$, which is kept fixed. Then, there is a real number $\varepsilon > 0$ such that (5.25) is convex in $\boldsymbol{\eta}^t$ on a convex neighborhood Θ^t for all $\boldsymbol{x}_t$ in the open ball $B_{2\varepsilon}(\boldsymbol{y}_t)$. A fortiori, the real-valued mapping

$$\kappa_t^{*-}(\boldsymbol{\eta}^t) := \frac{1}{\varepsilon}\Big[\langle \boldsymbol{c}_t^*(\boldsymbol{\eta}^t), \boldsymbol{y}_t \rangle + \alpha_t^{\text{o}}(\boldsymbol{\eta}^t)\Big]$$

is globally continuous and convex on a convex neighborhood of Θ^t. For the further argumentation we need the standard basis of $\mathbb{R}^{n_t}$, which is denoted by $\{\boldsymbol{e}_{t,i}\}_{i=1}^{n_t}$. Then, for every coordinate index $i = 1, \ldots, n_t$ the vector $\boldsymbol{y}_t + \varepsilon\, \boldsymbol{e}_{t,i}$ is an element of $B_{2\varepsilon}(\boldsymbol{y}_t)$. Consequently, the real function

$$\begin{aligned} \kappa_{t,i}^{*+}(\boldsymbol{\eta}^t) &:= \frac{1}{\varepsilon}\Big[\langle \boldsymbol{c}_t^*(\boldsymbol{\eta}^t), \boldsymbol{y}_t + \varepsilon\, \boldsymbol{e}_{t,i} \rangle + \alpha_t^{\text{o}}(\boldsymbol{\eta}^t)\Big] \\ &= c_{t,i}^*(\boldsymbol{\eta}^t) + \kappa_t^{*-}(\boldsymbol{\eta}^t) \end{aligned}$$

is globally continuous and convex on a convex neighborhood of Θ^t. Using the definitions

$$\boldsymbol{\kappa}_t^{*+} := (\kappa_{t,1}^{*+}, \ldots, \kappa_{t,n_t}^{*+}) \in \mathbb{R}^{n_t} \quad \text{and} \quad \boldsymbol{\kappa}_t^{*-} := (\kappa_t^{*-}, \ldots, \kappa_t^{*-}) \in \mathbb{R}^{n_t},$$

we can rewrite the vector of objective function coefficients as $\boldsymbol{c}_t^* = \boldsymbol{\kappa}_t^{*+} - \boldsymbol{\kappa}_t^{*-}$. This implies that $\boldsymbol{c}_t^*$ is d.c. on a convex neighborhood of Θ^t, and thus (F2) is established. In a next step we consider the constraints. By the regularity condition (E3), which is assumed to hold true in the present context, the constraint function (5.20) is regularizable in the sense of definition 5.7. Thus, proposition 5.8 guarantees that the involved recourse and technology matrices are constant, whereas the random vectors $\boldsymbol{h}_t^{\text{le}}$, $\boldsymbol{h}_t^{\text{ge}}$, and $\boldsymbol{h}_t^{\text{eq}}$ are d.c. on a convex neighborhood of Ξ^t. Finally, we have to prove that the recession cone $\{\boldsymbol{x}_t | W_t \boldsymbol{x}_t \sim_t \boldsymbol{0}\}$ of the feasible set $X_t(\boldsymbol{x}^{t-1}, \boldsymbol{\xi}^t)$ contains only the zero vector for all outcome and decision histories $(\boldsymbol{x}^{t-1}, \boldsymbol{\eta}^t, \boldsymbol{\xi}^t)$ in some neighborhood of Z^t. This follows immediately from boundedness of the feasible set mapping X_t on a neighborhood of Z^t. Consequently, (F3) is established. In summary, one may conclude that the linear stochastic program (5.18) complies with all of the regularity conditions (F1)–(F5). □

By proposition 5.21, the theoretical framework of Sect. 5.3 applies to any linear stochastic program of the form (5.18) which satisfies the regularity conditions (F1)–(F5). This implies, among other things, that the discrete barycentric measures P^l and P^u as well as the value functions Φ_t, Φ_t^l, and Φ_t^u are well-defined, and the static and dynamic versions of (5.18) are equivalent and solvable. Moreover, the optimal value of (5.18) has computationally accessible bounds, and the recourse functions Φ_t of the individual decision stages are squeezed between the auxiliary value functions Φ_t^l and Φ_t^u shifted by specific random variables. In order to construct these random variables, it is necessary to identify suitable correction terms of the form (5.15). To this end, we consider the dynamic version of the linear stochastic program (5.18). Then, the parametric stage t problem is given by

$$\begin{aligned} &\sup_{\boldsymbol{x}_t \in \mathbb{R}^{n_t}} \; q_t(\boldsymbol{x}^t, \boldsymbol{\eta}^t, \boldsymbol{\xi}^t) \\ &\text{s.t.} \quad W_t(\boldsymbol{\xi}^t)\boldsymbol{x}_t + T_t(\boldsymbol{\xi}^t)\boldsymbol{x}_{t-1} \sim_t \boldsymbol{h}_t(\boldsymbol{\xi}^t), \end{aligned} \tag{5.26}$$

where

$$q_t(\boldsymbol{x}^t, \boldsymbol{\eta}^t, \boldsymbol{\xi}^t) := \begin{cases} \langle \boldsymbol{c}_t^*(\boldsymbol{\eta}^t), \boldsymbol{x}_t \rangle + (E_t \Phi_{t+1})(\boldsymbol{x}^t, \boldsymbol{\eta}^t, \boldsymbol{\xi}^t) & \text{for } t < T, \\ \langle \boldsymbol{c}_T^*(\boldsymbol{\eta}^T), \boldsymbol{x}_T \rangle & \text{for } t = T. \end{cases}$$

Notice again that this representation explicitly deals with inequality (less-or-equal, greater-or-equal) and equality constraints. Let

$$X_{\text{opt},t}(\boldsymbol{x}^{t-1}, \boldsymbol{\eta}^t, \boldsymbol{\xi}^t) \subset \mathbb{R}^{n_t} \quad \text{and} \quad D_{\text{opt},t}^*(\boldsymbol{x}^{t-1}, \boldsymbol{\eta}^t, \boldsymbol{\xi}^t) \subset \mathbb{R}^{r_t^{\text{eff}}}$$

be the primal and dual solution sets associated with the parametric optimization problem (5.26), respectively. Based on the reasoning of Sect. 5.2 and before, one can show that the multifunctions $X_{\text{opt},t}$ and $D^*_{\text{opt},t}$ are nonempty-valued, bounded, and usc on a neighborhood of Z^t. Thus, there are nonnegative bounding vectors $\boldsymbol{X}_t^+, \boldsymbol{X}_t^- \in \mathbb{R}^{n_t}$ and $\boldsymbol{D}_t^{*+}, \boldsymbol{D}_t^{*-} \in \mathbb{R}^{r_t^{\text{eff}}}$ such that

$$-\boldsymbol{X}_t^- < \boldsymbol{x}_t < \boldsymbol{X}_t^+ \quad \forall \boldsymbol{x}_t \in X_{\text{opt},t}(Z^t) \tag{5.27a}$$

and

$$-\boldsymbol{D}_t^{*-} < \boldsymbol{d}_t^* < \boldsymbol{D}_t^{*+} \quad \forall \boldsymbol{d}_t^* \in D^*_{\text{opt},t}(Z^t). \tag{5.27b}$$

Without loss of generality, we may assume that (5.27a) not only holds for all optimal decisions in $X_{\text{opt},t}(Z^t)$ but also for all feasible decisions in $X_t(Z^t)$. Otherwise, one can add deterministic constraints to the parametric optimization problem (5.26) which guarantee the strict inequalities (5.27a) for all $\boldsymbol{x}_t \in X_t(Z^t)$ and which are non-binding at the optimum for any choice of the parameters in Z^t. Slackness implies that these additional constraints have no influence on the solution of (5.26) and may thus be incorporated without concern. By means of the bounding vectors for the primal and dual solutions it is possible to define appropriate *correction terms* α_t^{o} and α_t^{r} as in Sects. 5.1 and 5.2, respectively.

$$\alpha_t^{\text{o}}(\boldsymbol{\eta}^t) := +\langle \boldsymbol{\kappa}_t^{*-}(\boldsymbol{\eta}^t), \boldsymbol{X}_t^+ \rangle + \langle \boldsymbol{\kappa}_t^{*+}(\boldsymbol{\eta}^t), \boldsymbol{X}_t^- \rangle \tag{5.28a}$$

$$\alpha_t^{\text{r}}(\boldsymbol{\xi}^t) := -\langle \boldsymbol{D}_t^{*+}, \boldsymbol{\kappa}_t^+(\boldsymbol{\xi}^t) \rangle - \langle \boldsymbol{D}_t^{*-}, \boldsymbol{\kappa}_t^-(\boldsymbol{\xi}^t) \rangle \tag{5.28b}$$

These definitions reflect the intrinsic primal-dual symmetry of linear (stochastic) programs. Obviously, the correction term α_t^{o} associated with the nonconvexities in the objective function has the same general structure as the correction term α_t^{r} corresponding to the nonconvexities in the constraints. Concretely speaking, (5.28a) pairs the bounding vectors of the primal solutions with the d.c. components of the objective function coefficients, whereas (5.28b) pairs the bounding vectors of the dual solutions with the d.c. components of the rhs vector.

In proposition 5.21 we have shown that α_t^{o} is in fact a suitable correction term in the sense of definition 5.1 – remember that, without loss of generality, the estimate (5.27a) may be assumed to hold for all feasible decisions in $X_t(Z^t)$. Furthermore, it is obvious that α_t^{r} is of the form (5.6). Thus, it is a valid correction term in the sense of Sect. 5.2. Summing up both correction terms as in (5.15), one obtains an overall correction term $\alpha_t := \alpha_t^{\text{o}} + \alpha_t^{\text{r}}$. Next, introduce the conditional correction terms as in (5.16) and (5.17). Then, the conclusions of theorem 5.16 hold true, i.e. we have

$$\Phi_t^l + A_t^l - A_t \leq \Phi_t \leq \Phi_t^u + A_t^u - A_t \quad \text{on } Z^t \text{ for all } t \in \tau \tag{5.29a}$$

and

$$(E^l \Phi_0^l) + (E^l A_0^l) - (EA_0) \le (E\Phi_0) \le (E^u \Phi_0^u) + (E^u A_0^u) - (EA_0). \quad (5.29b)$$

By (5.28) the random variables A_t, A_t^l, and A_t^u are linear functionals of the bounding vectors for the primal and dual solution sets. Continuity then implies that (5.29) even holds if the strict inequalities in (5.27) are replaced by ordinary inequalities. Finally, it is also clear that the conclusions of theorem 5.17 remain valid, i.e. the upper and lower bounds on the recourse functions can be made arbitrarily tight by suitably refining the barycentric probability measures.

5.5 Bounding Sets for the Optimal Decisions

In Sect. 5.3 we found computable bounds bracketing the optimal value of a stochastic program which satisfies the regularity conditions (E1)–(E5). Recall that these conditions are weaker than any other set of regularity conditions considered in the present chapter. Below, we will argue that bounds on the optimal value also entail *bounding sets* for the optimal first stage decisions. As in Sect. 4.6, we explicitly work with the refinement parameter $J \in \mathbb{N}$ and use the standard assumption that the random outcomes at stage 0 are deterministic. Thus, we consider $(\boldsymbol{\eta}_0, \boldsymbol{\xi}_0)$ as a fixed parameter in the following exposition. Given any stochastic program subject to the conditions (E1)–(E5), we may introduce three extended-real-valued functionals

$$\begin{aligned}
F(\boldsymbol{x}_0) &:= \hat{\rho}_0(\boldsymbol{x}_0, \boldsymbol{\eta}_0, \boldsymbol{\xi}_0) + (E\,\Phi_1)(\boldsymbol{x}_0, \boldsymbol{\eta}_0, \boldsymbol{\xi}_0), \\
F_J^l(\boldsymbol{x}_0) &:= \hat{\rho}_0(\boldsymbol{x}_0, \boldsymbol{\eta}_0, \boldsymbol{\xi}_0) + (E_J^l \Phi_{1,J}^l)(\boldsymbol{x}_0, \boldsymbol{\eta}_0, \boldsymbol{\xi}_0) + A_{0,J}^l(\boldsymbol{\eta}_0, \boldsymbol{\xi}_0) - A_0(\boldsymbol{\eta}_0, \boldsymbol{\xi}_0), \\
F_J^u(\boldsymbol{x}_0) &:= \hat{\rho}_0(\boldsymbol{x}_0, \boldsymbol{\eta}_0, \boldsymbol{\xi}_0) + (E_J^u \Phi_{1,J}^u)(\boldsymbol{x}_0, \boldsymbol{\eta}_0, \boldsymbol{\xi}_0) + A_{0,J}^u(\boldsymbol{\eta}_0, \boldsymbol{\xi}_0) - A_0(\boldsymbol{\eta}_0, \boldsymbol{\xi}_0).
\end{aligned}$$

Then, the optimal value of the stochastic program under consideration is given by $\max F$, while the set of optimal first stage decisions coincides with $\arg\max F$. By theorem 5.16 we find $F_J^l \le F \le F_J^u$ on $\mathbb{R}^{n_0}$. Thus, the functional F is squeezed between the auxiliary functionals F_J^l and F_J^u on the entire underlying space. This observation implies that the maximizers of F are necessarily contained in the set of all $\boldsymbol{x}_0$ such that $F_J^u(\boldsymbol{x}_0)$ is greater or equal to the maximum of the lower bounding function F_J^l, i.e.

$$\arg\max F \subset C_J := \{\boldsymbol{x}_0 \,|\, F_J^u(\boldsymbol{x}_0) \ge \sup F_J^l\}.$$

Moreover, theorem 5.17 implies that the sequences $\{F_J^l\}_{J\in\mathbb{N}}$ and $\{F_J^u\}_{J\in\mathbb{N}}$ converge to F uniformly on its effective domain. Then, in analogy to theorem 4.9, we derive the following convergence result.

Theorem 5.22. *Consider a multistage stochastic program subject to the regularity conditions (E1)–(E5). Then, the sequence of upper level sets* $\{C_J\}_{J\in\mathbb{N}}$ *converges to* $\arg\max F$ *provided that the underlying refinement strategy is regular in the sense of (4.37).*

6

Applications in the Power Industry

Problems of power systems planning and operation as well as energy trading are often addressed with methods of stochastic programming. A literature review of relevant work in this field is provided in [106]. The existing (both static and dynamic) stochastic programming models can be classified according to their planning horizon. Long-term planning models deal with investments and typically have a horizon of up to 20 years. Medium-term planning is done over a range of 1 to 3 years and is concerned with resource management (mainly reservoir management). Short-term planning has a horizon of at most one week and typically deals with unit commitment and economic dispatch.

A second possibility to categorize energy models in stochastic programming arises from the ongoing deregulation of electricity markets. In fact, one may distinguish cost minimization models for an entire system and profit maximization models for individual agents. In a regulated market one usually adopts the perspective of a social planner. Then, the control objective is to meet demand at minimal cost with available generation equipment. In a liberalized market with a power exchange, however, it suffices to consider single agents, who choose their operating and trading decisions on the basis of price information. Consequently, it is not necessary to model the entire power system. All information about aggregate demand and supply as well as availability of resources and network congestions, which is relevant for an agent at a specific location, is contained in the local electricity prices. In other words, the local prices completely reflect the state of the system, and additional random variables become obsolete.

There exists a vast variety of solution methods to address the above optimization problems. For instance, *stochastic dynamic programming* (SDP) has been used for a long time to solve several types of energy models, cf. the surveys [101, 112, 113]. In SDP the dynamic program (3.2) is solved directly by backward recursion. Then, the value function is approximately evaluated at finitely many points and interpolated in between. Notice that this approach is only implementable if the value function has few arguments. Otherwise, computational effort explodes by the curse of dimensionality. The limitations of SDP are discussed in [82]. One solution method which overcomes some of the deficiencies of SDP is *stochastic dual dynamic programming* (SDDP) due to Pereira and Pinto [81–83] – see also the case study in [65]. This particular method is well suited for the solution of linear stochastic programs with non-anticipative constraint multifunction, provided that the expectation functional of any time stage is jointly concave in all of its arguments. Then, each expectation functional can be approximated by the lower envelope of finitely many hyperplanes. SDDP is mostly used for medium-term planning of regulated hydro-thermal power systems. For completeness, it is worthwhile to mention that SDP can principally be applied to any (potentially nonconvex) stochastic model, whereas the use of SDDP is limited to convex stochastic programs with a special dependence on the random parameters.

Short-term planning models are usually nonconvex as they include commitment decisions (binary 'on-off' decisions). Other integer variables might arise due to the precise modelling of nonlinear phenomena and technical restrictions. Stochastic models with commitment decisions are considered in [4, 86, 102–104]. However, for any reasonable discretization of the underlying probability space, the associated *mixed integer linear programs* (MILP) become too large to be solved directly – even in the two-stage case. The standard approach to overcome this difficulty involves *Lagrangian relaxation* of certain constraints and solution of the corresponding dual minimization problem (see e.g. [21] for an intuitive explanation of the Lagrangian relaxation approach). Thereby, one exploits the fact that the dual objective function frequently decomposes into small subproblems which can be tackled with a common MILP solver. Moreover, the dual objective function is convex in the Lagrange multipliers, and thus the standard tools of convex analysis can be used for its minimization (most authors favor specific *subgradient methods* or *proximal bundle methods* [37,54,63]). Unfortunately, due to the integrality restrictions there is a non-vanishing duality gap, which implies that the optimal value of the dual problem is merely an upper bound for the optimal value of the primal problem. Sometimes the duality gap can be shown to be small. Then, one attempts to find a nearly optimal primal solution by means of a suitable heuristics (see [22] for an analysis of the duality gaps corresponding to different relaxation schemes). In addition, the upper bounds obtained via dualization are often used in the bounding procedure of a *branch and bound algorithm* [15,77]; cf. also the slightly different approach in [68]. Notice that branch and bound algorithms principally find the primal optimum with arbitrary precision.

As pointed out by Römisch and Schultz [99], in typical energy models one can distinguish three types of constraints. *Dynamic constraints* describe the relation between decisions at different stages while *non-anticipativity constraints* couple the decisions associated with different scenarios. Furthermore, most energy models consist of lower-dimensional subproblems which are loosely coupled. These subordinate models are termed components. For ease of exposition, one can think of the components as models representing isolated plants in a larger power system. Then, the constraints relating decisions of different components are referred to as *component coupling constraints.* Lagrangian relaxation of dynamic, nonanticipativity, or component coupling constraints allows for nodal, scenario, or component decomposition of the dual objective function, respectively [99]. In [22, theorem 4.1] it is shown that scenario decomposition outperforms nodal decomposition as it spot maalways leads to a smaller duality gap. Thus, most authors focus on scenario or component decomposition (see e.g. [14] for a direct comparison).

Carøe and Schultz [15] propose a branch and bound algorithm based on scenario decomposition. This approach has successfully been applied to the solution of two-stage stochastic programs with integer requirements in the field of power production and trading [77]. On the other hand, Dentcheva, Groewe-Kuska, Römisch, et al. solve a multistage unit commitment problem by means of component decomposition [21, 49, 50, 73, 75, 76, 78]. Thereby, a suitable Lagrangian heuristics provides approximate solutions for the primal problem. A simplified model without commitment decisions is investigated in [98]. Furthermore, in [20] the direct solution by using a standard MILP solver is compared with the component decomposition method.

There are many other applications of stochastic programming in the energy industry. Among these, important contributions are due to Fleten, Wallace, and Ziemba, who present a portfolio model for a hydropower producer operating in a competitive electricity market [38, 39]. This nonlinear program accounts for energy generation and a set of power contracts for delivery and purchase, including contracts of financial nature. The authors do not report on numerical calculations, but they expect that Bender's decomposition method, SDDP, or a combination of different decomposition schemes will be best suited to solve their model. Moreover, Güssow [51] and Ostermaier [79] solve medium-term planning problems in a deregulated market by using the barycentric approximation scheme.

6.1 The Basic Decision Problem of a Hydropower Producer

Let us adopt the perspective of a hydropower producer in a liberalized electricity market. Assume that this producer has the right to operate one single pumped storage power plant over a finite planning horizon. First, choose a finite number of points in time (i.e. decision stages) indexed by $t \in \tau$, at which production and consumption decisions are selected. Define period t as the interval from time t to time $t+1$. Then, the stage t decision vector reads

$$\boldsymbol{x}_t := (x_{\text{sell},t}, x_{\text{buy},t}, x_{\text{store},t}),$$

and its components are measured in MW h. $x_{\text{sell},t}$ denotes the energy generated in period t while $x_{\text{buy},t}$ stands for the energy purchased on the rket to pump water into the reservoir. For simplicity, all nonlinear effects due to water level fluctuations are neglected. Instead, it is assumed that the water level is constant, and the amount of energy stored at the end of period t, i.e. $x_{\text{store},t}$, is proportional to the volume of water contained in the reservoir. Concretely speaking, $x_{\text{store},t}$ coincides with the potential energy of the stored water multiplied by the generation efficiency.

Using the above definitions, the owner of a pumped hydropower plant faces the following decision problem.

$$\max_{\boldsymbol{x}\in\mathcal{N}_{3(T+1)}} \int_{\Theta\times\Xi} \Big(\sum_{t=0}^{T} \eta_t\,(x_{\text{sell},t} - x_{\text{buy},t})\Big) dP(\boldsymbol{\eta},\boldsymbol{\xi}) \tag{6.1a}$$

$$\text{s.t.} \quad x_{\text{store},t} - x_{\text{store},t-1} + x_{\text{sell},t} - \epsilon_{\text{p}}\, x_{\text{buy},t} - \xi_t = 0 \quad \forall t \in \mathcal{T} \tag{6.1b}$$

$$\left.\begin{array}{lll} 0 & \le x_{\text{sell},t} & \le \overline{x}_{\text{sell},t} \\ 0 & \le x_{\text{buy},t} & \le \overline{x}_{\text{buy},t} \\ \underline{x}_{\text{store},t} & \le x_{\text{store},t} & \le \overline{x}_{\text{store},t} \end{array}\right\} \quad \forall t \in \mathcal{T} \tag{6.1c}$$

Thus, the power producer maximizes the expected revenues from electricity generation subject to the *energy balance equation* (6.1b) and the *capacity constraints* (6.1c). As usual, the constraints are assumed to hold almost surely with respect to the probability measure P. However, the attribute 'P-a.s'. is suppressed throughout this chapter in order to keep notation simple. η_t stands for the local electricity price, which is completely exogenous and can not be influenced by the power provider. On the other hand, ξ_t characterizes the amount of energy by which the reservoir content increases in period t due to natural water inflows. As outlined in the introduction, we assume that a producer in a perfectly liberalized market must not care about load demand and possible network congestions, since market clearing and stability of the transmission grid are regulated via local spot prices. Thus, we need no additional random parameters or supplementary constraints. For ease of exposition, suppose that the real-valued random variables η_t and ξ_t follow first order autoregressive processes with correlated noise.

$$\eta_t = H_t^{\text{o}}\, \eta_{t-1} + \varepsilon_t^{\text{o}}, \qquad \xi_t = H_t^{\text{r}}\, \xi_{t-1} + \varepsilon_t^{\text{r}} \tag{6.1d}$$

Notice that this specification covers the customary mean reversion processes. Besides the AR(1) coefficients H_t^{o} and H_t^{r}, the distributional parameters of the serially independent disturbances $(\varepsilon_t^{\text{o}}, \varepsilon_t^{\text{r}})$ must be specified. We assume that $(\varepsilon_t^{\text{o}}, \varepsilon_t^{\text{r}})$ follows a bivariate normal distribution *truncated* outside some regular cross-simplex, which is large enough to contain all the data with high probability. The underlying normal distribution has mean $\boldsymbol{\mu}_t$ and covariance matrix Σ_t.

$$\boldsymbol{\mu}_t := \begin{pmatrix} \mu_t^{\mathrm{o}} \\ \mu_t^{\mathrm{r}} \end{pmatrix}, \qquad \Sigma_t := \begin{pmatrix} (\sigma_t^{\mathrm{o}})^2 & \rho_t \sigma_t^{\mathrm{o}} \sigma_t^{\mathrm{r}} \\ \rho_t \sigma_t^{\mathrm{o}} \sigma_t^{\mathrm{r}} & (\sigma_t^{\mathrm{r}})^2 \end{pmatrix}$$

Any confidence level $\alpha_{\mathrm{c}} \gtrsim 0$ defines a *confidence ellipse* in the $(\varepsilon_t^{\mathrm{o}}, \varepsilon_t^{\mathrm{r}})$-space as the smallest Borel set containing the fraction $1-\alpha_{\mathrm{c}}$ of the probability mass of the underlying normal distribution. Then, choose the cross-simplex $\mathcal{E}_t^{\mathrm{o}} \times \mathcal{E}_t^{\mathrm{r}}$ to be the smallest two-dimensional interval covering the $1-\alpha_{\mathrm{c}}$ confidence ellipse. Finally, the distribution of the disturbances $(\varepsilon_t^{\mathrm{o}}, \varepsilon_t^{\mathrm{r}})$ is obtained via truncation of the normal distribution on the complement of $\mathcal{E}_t^{\mathrm{o}} \times \mathcal{E}_t^{\mathrm{r}}$ and normalization. An elementary calculation shows

$$\mathcal{E}_t^{\mathrm{o}} = \left[\mu_t^{\mathrm{o}} - \sigma_t^{\mathrm{o}} n_{\mathrm{c}},\, \mu_t^{\mathrm{o}} + \sigma_t^{\mathrm{o}} n_{\mathrm{c}}\right] \quad \text{and} \quad \mathcal{E}_t^{\mathrm{r}} = \left[\mu_t^{\mathrm{r}} - \sigma_t^{\mathrm{r}} n_{\mathrm{c}},\, \mu_t^{\mathrm{r}} + \sigma_t^{\mathrm{r}} n_{\mathrm{c}}\right], \tag{6.2}$$

where $n_{\mathrm{c}} := (\ln \alpha_{\mathrm{c}}^{-2})^{-1/2}$. Truncation of the normal distribution is necessary since extreme inflow scenarios could interfere with the requirement for nonanticipativity of the constraint multifunction. Without truncation, the natural inflows would exceed $\overline{x}_{\mathrm{store},t} + \overline{x}_{\mathrm{sell},t}$ with nonzero probability. This in turn implies that the energy balance equation (6.1b) could not be satisfied with probability 1. A second reason for truncation is due to the fact that negative spot prices or inflows lack physical meaning and should therefore be excluded. Last but not least, remember that the barycentric approximation scheme explicitly requires compact probability spaces.

Table 6.1. Input parameters

Param.	Description	Unit
$\overline{x}_{\mathrm{sell},t}$	max. amount of energy sold on the spot market	(MW h)
$\overline{x}_{\mathrm{buy},t}$	max. amount of energy bought on the spot market	(MW h)
$\overline{x}_{\mathrm{store},t}$	max. reservoir content (reservoir capacity)	(MW h)
$\underline{x}_{\mathrm{store},t}$	min. reservoir content (e.g. predetermined target value)	(MW h)
$x_{\mathrm{store},-1}$	reservoir content at the beginning of period 0	(MW h)
ϵ_{P}	energy conversion factor	(–)
H_t^{o}	AR(1) coefficient of the spot price process	(–)
H_t^{r}	AR(1) coefficient of the natural inflow process;	(–)
η_0	(deterministic) electricity spot price in the first period	(€ MW^{-1}h^{-1})
ξ_0	(deterministic) reservoir inflow in the first period	(MW h)
μ_t^{o}	mean value of $\varepsilon_t^{\mathrm{o}}$ without truncation	(€ MW^{-1}h^{-1})
μ_t^{r}	mean value of $\varepsilon_t^{\mathrm{r}}$ without truncation	(MW h)
σ_t^{o}	standard deviation of $\varepsilon_t^{\mathrm{o}}$ without truncation	(€ MW^{-1}h^{-1})
σ_t^{r}	standard deviation of $\varepsilon_t^{\mathrm{r}}$ without truncation	(MW h)
ρ_t	correlation coefficient of $\varepsilon_t^{\mathrm{o}}$ and $\varepsilon_t^{\mathrm{r}}$ without truncation	(–)
α_{c}	confidence level for truncation	(–)

With the aid of (6.2) we can in fact determine the marginal spaces Θ_t and Ξ_t covering the support of the spot price and the natural inflow in period t, respectively. To this end, rewrite η_t and ξ_t as a linear combination of the serially uncorrelated noise terms, i.e.

$$\eta_t = H^{\mathrm{o}}_{1,t}\eta_0 + \sum_{s=1}^{t} H^{\mathrm{o}}_{s+1,t}\varepsilon^{\mathrm{o}}_s, \quad \text{where} \quad H^{\mathrm{o}}_{s,t} := \begin{cases} \prod_{s'=s}^{t} H^{\mathrm{o}}_{s'} & \text{for } s \leq t, \\ 1 & \text{else,} \end{cases}$$

and

$$\xi_t = H^{\mathrm{r}}_{1,t}\xi_0 + \sum_{s=1}^{t} H^{\mathrm{r}}_{s+1,t}\varepsilon^{\mathrm{r}}_s, \quad \text{where} \quad H^{\mathrm{r}}_{s,t} := \begin{cases} \prod_{s'=s}^{t} H^{\mathrm{r}}_{s'} & \text{for } s \leq t, \\ 1 & \text{else.} \end{cases}$$

Under the reasonable assumption that the AR(1) coefficients in (6.1d) are nonnegative, we obtain

$$\Theta_t = \Big[H^{\mathrm{o}}_{1,t}\eta_0 + \sum_{s=1}^{t} H^{\mathrm{o}}_{s+1,t}(\mu^{\mathrm{o}}_s - \sigma^{\mathrm{o}}_s n_{\mathrm{c}}),\ H^{\mathrm{o}}_{1,t}\eta_0 + \sum_{s=1}^{t} H^{\mathrm{o}}_{s+1,t}(\mu^{\mathrm{o}}_s + \sigma^{\mathrm{o}}_s n_{\mathrm{c}})\Big],$$

$$\Xi_t = \Big[H^{\mathrm{r}}_{1,t}\xi_0 + \sum_{s=1}^{t} H^{\mathrm{r}}_{s+1,t}(\mu^{\mathrm{r}}_s - \sigma^{\mathrm{r}}_s n_{\mathrm{c}}),\ H^{\mathrm{r}}_{1,t}\xi_0 + \sum_{s=1}^{t} H^{\mathrm{r}}_{s+1,t}(\mu^{\mathrm{r}}_s + \sigma^{\mathrm{r}}_s n_{\mathrm{c}})\Big].$$

The relevant input parameters for the optimization problem (6.1) are listed in Table 6.1. It is worthwhile to remark that the *energy conversion factor* $\epsilon_{\mathrm{P}} < 1$ can be expressed as the pumping efficiency multiplied by the generating efficiency and accounts for the fact that electrical energy is not storable. In other words, only the fraction ϵ_{P} of a certain amount of energy bought on the spot market can later be reused. The rest is dissipated in the equipment of the power plant. For real pumped storage power plants the conversion factor ϵ_{P} approximately amounts to 75%.

It can easily be verified that the linear stochastic program (6.1) satisfies the regularity conditions (B1)–(B4). Moreover, assumption (B5) holds if the reservoir targets $\overline{x}_{\text{store},t}$ and $\underline{x}_{\text{store},t}$ as well as the confidence level α_{c} for truncation of the normal distribution are chosen appropriately, which will be implicitly assumed in the following. These conditions guarantee that the barycentric approximation scheme provides arbitrarily tight upper and lower bounds on the optimal value. In the sequel, we develop specific generalizations of the basic model (6.1).

6.2 Market Power

In Sect. 6.1 it is assumed that the amount of energy supplied (or bought) by the hydro generator does not influence the spot price. Thus, in a perfectly competitive energy market the spot price behaves like an exogenous stochastic process, implying that the generator has no market power. In an oligopolistic market, however, the price-taking assumption should be relaxed. Let us therefore assume that the electricity spot price is a monotonously decreasing linear affine function of the net energy production $x_{\text{net},t} := x_{\text{sell},t} - x_{\text{buy},t}$.

$$S_t(x_{\text{net},t}) := \eta_t\,(1 - c_t\, x_{\text{net},t})$$

Here, S_t denotes the spot price in period t contingent on the trading volume of electric energy whereas the random variable η_t characterizes a reference price, which is realized if the hydro power plant under consideration is shut down. This approach is consistent with the model developed in [5]. Using the terminology of microeconomics, S_t may be interpreted as an *inverse demand function* in an oligopolistic market [71, Chap. 12]. In fact, the output levels of the competitors can be viewed as exogenous random parameters which are implicitly taken account of in the reference price η_t. The nonnegative constant c_t is measured in $(\text{MW h})^{-1}$ and determines the *inverse demand elasticity* with respect to net production. It is assumed that $c_t^{-1} > \max\{\overline{x}_{\text{sell},t}, \overline{x}_{\text{buy},t}\}$ such that the spot price remains positive for all feasible production and consumption decisions. Plugging S_t into the objective function (6.1a) we obtain

$$\int_{\Theta\times\Xi}\Big(\sum_{t=0}^{T} S_t(x_{\text{net},t})\, x_{\text{net},t}\Big)\,dP(\boldsymbol{\eta},\boldsymbol{\xi}). \tag{6.3}$$

Since $c_t \geq 0$ we find that

$$\rho_t(x_{\text{net},t}, \eta_t) = \eta_t\,(x_{\text{net},t} - c_t\, x_{\text{net},t}^2) =: \eta_t\, f_t(x_{\text{net},t})$$

is a continuous saddle function being concave (quadratic) in $x_{\text{net},t}$ and convex (linear) in η_t on its natural domain. Due to the quadratic terms in the objective function, the hydro scheduling problem with market power does not represent a linear stochastic program. However, the conditions (B1)–(B5) remain valid, and the barycentric approximation scheme is applicable. In order to solve the discretized auxiliary stochastic programs (4.40) by means of an LP solver, the quadratic terms in (6.3) must be approximated by piecewise linear functions. To this end, we introduce a dummy variable $x_{\text{aux},t}$ and write

$$\begin{aligned} f_t(x_{\text{net},t}) &= \sup\,\{x_{\text{aux},t}\,|\,x_{\text{aux},t} \leq f_t(x_{\text{net},t})\} \\ &\approx \sup\,\{x_{\text{aux},t}\,|\,x_{\text{aux},t} \leq f_{t,J,i} + f'_{t,J,i}\, x_{\text{net},t},\ i \in \mathcal{I}_J\} \\ &=: f_{t,J}(x_{\text{net},t}), \end{aligned} \tag{6.4}$$

where $\mathcal{I}_J := \{1, \ldots, I(J)\}$ is a finite index set, and

$$f_{t,J,i},\ f'_{t,J,i} \in \mathbb{R} \quad \forall t \in \tau,\ J \in \mathbb{N},\ i \in \mathcal{I}_J.$$

In other words, $f_{t,J}$ is given by the lower envelope of the linear affine functions $f_{t,J,i} + f'_{t,J,i}\, x_{\text{net},t}$, $i \in \mathcal{I}_J$. The parameter J labels the different piecewise linear approximations $f_{t,J}$ of the quadratic function f_t. In the remainder, let us assume that the sequence $\{f_{t,J}\}_{J\in\mathbb{N}}$ converges to f_t uniformly on the compact interval $[-\overline{x}_{\mathrm{buy},t}, \overline{x}_{\mathrm{sell},t}]$.

Denote by $(E^l\Phi^l_{0,J})$ and $(E^u\Phi^u_{0,J})$ the optimal values of the hydro scheduling problem with market power if the quadratic functions f_t are approximated by the piecewise linear functions $f_{t,J}$ and the original measure P is substituted by the discrete barycentric measures P^l and P^u, respectively.[1] Notice that the difference between these optimal values can be made small by suitably refining the barycentric probability measures. Next, for a given tolerance $\varepsilon > 0$ there is a parameter $J_0(\varepsilon)$ such that

$$|f_{t,J} - f_t| \leq \frac{\varepsilon}{(T+1)\,E(\eta_t)} \quad \forall t \in \tau,\ J \geq J_0(\varepsilon) \tag{6.5}$$

uniformly on $[-\overline{x}_{\mathrm{buy},t}, \overline{x}_{\mathrm{sell},t}]$. Then, theorem 4.7 together with an elementary stability result implies that

$$(E^l\Phi^l_{0,J}) - \varepsilon \leq (E^l\Phi^l_0) \leq (E\Phi_0) \leq (E^u\Phi^u_0) \leq (E^u\Phi^u_{0,J}) + \varepsilon.$$

Thus, the optimal value $(E\Phi_0)$ of the hydro scheduling problem with market power has upper and lower bounds which are computationally accessible by means of an LP solver. Moreover, the bounds can be made arbitrarily tight by refining the discrete barycentric measures and by improving the piecewise linear approximations of the quadratic terms in the objective function.

6.3 Lognormal Spot Prices

In the basic model (6.1) it is assumed that the spot price process is governed by a simple Brownian motion with mean reversion. As mentioned above, this specification allows the spot price to drop below zero if the involved normal distributions are not suitably truncated. Moreover, the magnitude of the price fluctuations is independent of the current price level. However, actual electricity prices are nonnegative, and price fluctuations are expected to be higher when the price level is high. Geometric Brownian motion exhibits exactly these characteristics, and thus it is more adequate for modelling spot prices. Subsequently, we assume that the spot price at time t is lognormally distributed, $S_t := \exp(\eta_t)$, where η_t follows an AR(1) process of the form (6.1d).[2] Then, the objective function can be rewritten as

[1] Here, the refinement parameter corresponding to the barycentric measures is suppressed. Otherwise, in order to avoid any ambiguity, the refinement parameter and the parameter labelling the piecewise linearization should be denoted by J_1 and J_2, respectively.

[2] Lognormal electricity prices with mean reversion have previously been studied by Deng et al. [19].

$$\int_{\Theta \times \Xi} \Big(\sum_{t=0}^{T} \exp(\eta_t)\, x_{\text{net},t} \Big) dP(\boldsymbol{\eta}, \boldsymbol{\xi}), \tag{6.6}$$

where $x_{\text{net},t} := x_{\text{sell},t} - x_{\text{buy},t}$ as in the previous section. Unfortunately, the resulting stochastic program does not fulfill the regularity conditions (B1)–(B5). In fact, for $x_{\text{net},t} < 0$ the stage t profit function is concave in η_t, which contradicts condition (B2). Consequently, the auxiliary value functions do not necessarily provide bounds for the optimal value of the recourse problem under consideration. However, it can easily be verified that the hydro scheduling problem with lognormal spot prices represents a linear stochastic program subject to the generalized regularity conditions (C1)–(C5).[3] In order to derive bounds on its optimal value, we first have to identify suitable correction terms $\{\alpha_t\}_{t \in \mathcal{T}}$ in the sense of definition 5.1. Since η_0 is deterministic, a possible choice is

$$\alpha_t(\eta_t) := \begin{cases} X^-_{\text{net},t} \exp(\eta_t) & \text{for } t = 1, \ldots, T, \\ 0 & \text{for } t = 0, \end{cases}$$

with $X^-_{\text{net},t} > \overline{x}_{\text{buy},t}$ arbitrary. According to theorem 5.5, the bounds on $(E\Phi_0)$ consist in the auxiliary value functions of the barycentric approximation scheme shifted by a combination of the expectation values (EA_0), $(E^l A^l_0)$, and $(E^u A^u_0)$. The auxiliary value functions as well as $(E^l A^l_0)$ and $(E^u A^u_0)$ are conveniently evaluated numerically. Moreover, notice that the expectation (EA_0) with respect to the truncated normal distribution is analytically untractable. One can either evaluate (EA_0) by using numerical integration, or – provided that the confidence level α_c is small enough – one may neglect truncation and calculate the expectation with respect to the unrestricted normal distribution. To this end, rewrite η_t as a linear combination of the noise terms, i.e.

$$\eta_t = H^{\text{o}}_{1,t}\eta_0 + \sum_{s=1}^{t} H^{\text{o}}_{s+1,t}\varepsilon^{\text{o}}_s, \quad \text{where} \quad H^{\text{o}}_{s,t} := \begin{cases} \prod_{s'=s}^{t} H^{\text{o}}_{s'} & \text{for } s \le t, \\ 1 & \text{else.} \end{cases}$$

Then, disregarding truncation, we obtain

$$\begin{aligned} (EA_0) &= \sum_{t=1}^{T} X^-_{\text{net},t}\, E(\exp(\eta_t)) \\ &\approx \sum_{t=1}^{T} X^-_{\text{net},t} \exp\Big(H^{\text{o}}_{1,t}\eta_0 + \sum_{s=1}^{t} H^{\text{o}}_{s+1,t}\mu^{\text{o}}_s + \frac{1}{2}(H^{\text{o}}_{s+1,t}\sigma^{\text{o}}_s)^2 \Big). \end{aligned}$$

6.4 Lognormal Natural Inflows

For similar reasons as in the case of the spot prices, it is meaningful to work with lognormally distributed reservoir inflows. Let us therefore assume that

[3] Of course, we could equivalently invoke the regularity conditions (F1)–(F5).

the natural inflows in period t are given by $\exp(\xi_t)$, where ξ_t follows an AR(1) process of the form (6.1d). This assumption requires a modification of the energy balance equation.

$$x_{\text{store},t} - x_{\text{store},t-1} + x_{\text{sell},t} - \epsilon_{\text{P}}\, x_{\text{buy},t} - \exp(\xi_t) = 0 \tag{6.7}$$

It is straightforward to verify that the corresponding constraint function is nonlinear in ξ_t – an obvious violation of condition (B3). We may thus conclude that the auxiliary value functions do not provide strict bounds on the recourse functions of the underlying stochastic program. All the same, one can verify that the hydro scheduling problem with lognormal inflows fulfills the generalized regularity conditions (D1)–(D5).[4] In order to employ theorem 5.14, let us look for a sequence of correction terms $\{\alpha_t\}_{t\in\mathcal{T}}$ as in (5.6). Thus, study the dynamic version of the hydro scheduling problem at hand. Restricting attention to the parametric stage t subproblem, we should find an upper bound $D^{*+}_{\text{bal},t}$ for the dual variable associated with the energy balance equation (6.7) uniformly over all outcome and decision histories in Z^t. In economic literature, this dual variable is usually referred to as the *shadow price* of energy. As a matter of fact, it can be interpreted as the maximum price the hydropower producer would be ready to pay at time t for the use of a small additional amount of energy $\Delta x_{\text{store},t-1}$. This observation implies that the dual variable of restriction (6.7) is nonnegative since more energy in the reservoir necessarily results in higher profits. Without loss of generality, we may therefore interpret the energy balance equation (6.7) as a 'less-or-equal' constraint. A small additional amount of energy $\Delta x_{\text{store},t-1}$ in the reservoir can either be sold on the spot market or – in times of scarcity – it can be used to meet the reservoir targets in the subsequent periods. In the former case, selling the extra amount of energy on the spot market results in an excess profit of at most $\eta_{\max,t,T}\,\Delta x_{\text{store},t-1}$, where

$$\eta_{\max,t,T} := \max\{\,\sup\{\eta_s \in \Theta_s\} \mid s = t,\ldots,T\}.$$

In the latter case, the hydropower producer saves at most a lump sum of $(\eta_{\max,t,T}/\epsilon_{\text{P}})\,\Delta x_{\text{store},t-1}$. Notice that $\eta_{\max,t,T}$ tends to infinity as the confidence level for truncation becomes small. Under the given assumptions, $\eta_{\max,t,T}$ can be calculated analytically, i.e. it can be expressed as a function of the basic input parameters in Table 6.1. By the above reasoning, any real number $D^{*+}_{\text{bal},t} > \eta_{\max,t,T}/\epsilon_{\text{P}}$ represents a strict uniform upper bound for the shadow price of energy at time t. Being aware of this result, we are now prepared to define correction terms as in (5.6). Since the random variable ξ_0 is deterministic, a suitable choice is

$$\alpha_t(\xi_t) := \begin{cases} -D^{*+}_{\text{bal},t}\,\exp(\xi_t) & \text{for } t = 1,\ldots,T, \\ 0 & \text{for } t = 0. \end{cases}$$

[4] Here again, we could equivalently work with the regularity conditions (F1)–(F5) for linear stochastic programs.

According to theorem 5.14, the bounds on $(E\Phi_0)$ consist in the auxiliary value functions of the barycentric approximation scheme shifted by a combination of the expectation values (EA_0), $(E^l A_0^l)$, and $(E^u A_0^u)$. As already mentioned in the previous section, the auxiliary value functions as well as $(E^l A_0^l)$ and $(E^u A_0^u)$ can be treated numerically. In addition, (EA_0) can either be evaluated by using numerical integration or by neglecting truncation and calculating the expectation with respect to the unrestricted normal distribution. To this end, rewrite ξ_t as a linear combination of the noise terms, i.e.

$$\xi_t = H^{\mathrm{r}}_{1,t}\xi_0 + \sum_{s=1}^{t} H^{\mathrm{r}}_{s+1,t}\varepsilon^{\mathrm{r}}_s, \quad \text{where} \quad H^{\mathrm{r}}_{s,t} := \begin{cases} \prod_{s'=s}^{t} H^{\mathrm{r}}_{s'} & \text{for } s \leq t, \\ 1 & \text{else.} \end{cases}$$

Then, in analogy to the previous section we find

$$\begin{aligned} (EA_0) &= -\sum_{t=1}^{T} D^{*+}_{\mathrm{bal},t}\, E(\exp(\xi_t)) \\ &\approx -\sum_{t=1}^{T} D^{*+}_{\mathrm{bal},t} \exp\Big(H^{\mathrm{r}}_{1,t}\xi_0 + \sum_{s=1}^{t} H^{\mathrm{r}}_{s+1,t}\mu^{\mathrm{r}}_s + \frac{1}{2}(H^{\mathrm{r}}_{s+1,t}\sigma^{\mathrm{r}}_s)^2\Big). \end{aligned}$$

6.5 Risk Aversion

A more general objective function than (6.1a) is

$$\int_{\Theta\times\Xi} U\Big(\sum_{t=0}^{T} \eta_t\, x_{\mathrm{net},t}\Big) dP(\boldsymbol{\eta}, \boldsymbol{\xi}), \tag{6.8}$$

where $U : \mathbb{R} \to \mathbb{R}$ is a *utility function* of von Neumann-Morgenstern type. Here, U is twice continuously differentiable and concave so as to reflect *risk aversion.*[5] The standard approach for firms to manage risk is not to use concave utility functions, but mitigate risk through net present value calculations under a risk-adjusted probability measure. However, the expected utility approach is adequate if a firm has only few owners with limited opportunities to diversify. In order to keep notation simple, set $x_{\mathrm{net},t} := x_{\mathrm{sell},t} - x_{\mathrm{buy},t}$ for all $t \in \tau$ and $\boldsymbol{x}^T_{\mathrm{net}} := (x_{\mathrm{net},0}, \ldots, x_{\mathrm{net},T})$. These definitions are consistent with previous conventions. Apparently, the objective function (6.8) can be interpreted as the expected value of a sum of profit functions

[5]Utility functions of wealth accumulated over the entire planning horizon are e.g. used in [1]. However, the present exposition could also be based on more general utility functions as in [38], provided that they are concave in the decision variables and sufficiently smooth. For instance, consumption-oriented models usually invoke time-separable utility.

$$\rho_t(\boldsymbol{x}^t, \boldsymbol{\eta}^t) = \begin{cases} U(\langle \boldsymbol{\eta}^T, \boldsymbol{x}^T_{\text{net}} \rangle) & \text{for } t = T, \\ 0 & \text{for } t = 0, \ldots, T-1. \end{cases}$$

This convention fits into the formal framework of Sect. 2.4. It can easily be checked that ρ_T is concave in $\boldsymbol{x}^T$ for every fixed value of $\boldsymbol{\eta}^T$ and concave in $\boldsymbol{\eta}^T$ for all fixed values of $\boldsymbol{x}^T$. In other words, ρ_T is biconcave (but not jointly concave) in $\boldsymbol{\eta}^T$ and $\boldsymbol{x}^T$ implying that condition (B2) is not fulfilled. Hence, the auxiliary value functions of barycentric approximation do not furnish bounds on the optimal objective function value. However, the hydro scheduling problem with risk aversion represents a convex stochastic program which satisfies the generalized regularity conditions (C1)–(C5). This is a direct consequence of proposition 5.3. In order to find suitable correction terms $\{\alpha_t\}_{t\in\mathcal{T}}$ as in definition 5.1, let us estimate the curvature of the profit function ρ_T. An elementary calculation shows (cf. also proposition 5.3)

$$\|\nabla_{\boldsymbol{\eta}^T} \otimes \nabla_{\boldsymbol{\eta}^T} \rho_T(\boldsymbol{x}^T, \boldsymbol{\eta}^T)\|_2 = \|\boldsymbol{x}^T_{\text{net}}\|^2 \, |U''(\langle \boldsymbol{\eta}^T, \boldsymbol{x}^T_{\text{net}} \rangle)|,$$

where U'' stands for the second derivative of the utility function U. Then, any strictly positive real number

$$c_T^2 > \sup\left\{ \|\boldsymbol{x}^T_{\text{net}}\|^2 \, |U''(\langle \boldsymbol{\eta}^T, \boldsymbol{x}^T_{\text{net}} \rangle)| \;\middle|\; (\boldsymbol{x}^T, \boldsymbol{\eta}^T, \boldsymbol{\xi}^T) \in Y^T \right\}$$

represents an upper bound for the curvature of ρ_T along $\boldsymbol{\eta}^T$ on a neighborhood of Y^T. Thus, a possible choice for the correction terms is

$$\alpha_t(\eta_t) := \begin{cases} c_T^2 \, \langle \boldsymbol{\eta}^T, \boldsymbol{\eta}^T \rangle & \text{for } t = T, \\ 0 & \text{for } t = 0, \ldots, T-1. \end{cases}$$

As implied by theorem 5.5, the bounds on $(E\Phi_0)$ are made up of the auxiliary value functions and the expectation values (EA_0), $(E^l A_0^l)$, and $(E^u A_0^u)$. The numerical evaluation of the auxiliary value functions as well as $(E^l A_0^l)$ and $(E^u A_0^u)$ poses no severe difficulties. In contrast, the expectation (EA_0) with respect to the truncated normal distribution is analytically untractable. Thus, one can attempt to evaluate (EA_0) by using numerical integration. Alternatively, one may neglect truncation and calculate the expectation with respect to the unrestricted normal distribution. To this end, express η_t as a linear combination of the noise terms, as exemplified in Sect. 6.3. Then, disregarding truncation, we obtain

$$\begin{aligned} (EA_0) &= E(\alpha_T(\boldsymbol{\eta}^T)) \\ &\approx c_T^2 \sum_{t=0}^{T} \left[\left(H^{\circ}_{1,t}\eta_0 + \sum_{s=1}^{t} H^{\circ}_{s+1,t}\mu^{\circ}_s \right)^2 + \sum_{s=1}^{t} \left(H^{\circ}_{s+1,t}\sigma^{\circ}_s \right)^2 \right]. \end{aligned}$$

Notice that the hydro scheduling problem with risk aversion represents a convex stochastic program with nonlinear objective. In order to solve the associated auxiliary stochastic programs (4.40) by means of an LP solver, the nonlinear terms in (6.8) must be approximated by piecewise linear functions. To this end, it is useful to introduce an additional decision variable $x_{\text{rev},t}$ for the total revenues earned until the end of period t. The evolvement of $x_{\text{rev},t}$ is determined by the *revenue balance equation*

$$x_{\text{rev},t} - x_{\text{rev},t-1} - \eta_t \, x_{\text{net},t} = 0 \quad \forall t \in \tau, \tag{6.9}$$

and initial wealth $x_{\text{rev},-1}$ is set to zero. For the further argumentation we need rough upper and lower estimates for the terminal wealth $x_{\text{rev},T}$ uniformly over all outcomes and all decision strategies feasible in (4.40). A suitable choice is

$$\left.\begin{aligned} \underline{x}_{\text{rev},T} &:= -\textstyle\sum_{t=0}^{T} \eta_{\max,t}\, \overline{x}_{\text{buy},t} \\ \overline{x}_{\text{rev},T} &:= \textstyle\sum_{t=0}^{T} \eta_{\max,t}\, \overline{x}_{\text{sell},t} \end{aligned}\right\} \quad \text{where} \quad \eta_{\max,t} := \sup\{\eta_t \in \Theta_t\}.$$

With these conventions, the stage T profit function reduces to $U(x_{\text{rev},T})$ and can conveniently be approximated by a piecewise linear function.

$$\begin{aligned} U(x_{\text{rev},T}) &= \sup\,\{x_{\text{aux},T} \,|\, x_{\text{aux},T} \le U(x_{\text{rev},T})\} \\ &\approx \sup\,\{x_{\text{aux},T} \,|\, x_{\text{aux},T} \le U_{J,i} + U'_{J,i}\, x_{\text{rev},T},\; i \in \mathcal{I}_J\} \\ &=: U_J(x_{\text{rev},T}) \end{aligned} \tag{6.10}$$

Here, $x_{\text{aux},t}$ is a dummy variable, $\mathcal{I}_J := \{1,\ldots,I(J)\}$ is a finite index set, and

$$U_{J,i},\, U'_{J,i} \in \mathbb{R} \quad \forall J \in \mathbb{N},\, i \in \mathcal{I}_J.$$

Explicitly speaking, U_J is given by the lower envelope of the linear affine functions $U_{J,i} + U'_{J,i}\, x_{\text{rev},T}$, $i \in \mathcal{I}_J$. The parameter J labels the different piecewise linear approximations U_J of the concave utility function U. In the remainder, let us assume that the sequence $\{U_J\}_{J\in\mathbb{N}}$ converges to U uniformly on the compact interval $[\underline{x}_{\text{rev},T}, \overline{x}_{\text{rev},T}]$.

In analogy to Sect. 6.2, denote by $(E^l\Phi^l_{0,J})$ and $(E^u\Phi^u_{0,J})$ the optimal values of the hydro scheduling problem with risk aversion if the nonlinear utility function U is approximated by the piecewise linear function U_J, and the original measure P is substituted by the barycentric measures P^l and P^u, respectively.[6] Next, for a given tolerance $\varepsilon > 0$ there is a parameter $J_0(\varepsilon)$ such that

$$|U_J - U| \le \varepsilon \quad \forall J \ge J_0(\varepsilon)$$

uniformly on $[\underline{x}_{\text{rev},T}, \overline{x}_{\text{rev},T}]$. Then, theorem 5.5 together with an elementary stability result implies that

[6] As usual, the refinement parameter corresponding to the barycentric measures is suppressed. Otherwise, for the sake of consistent notation, the refinement parameter and the parameter labelling the piecewise linearization should be denoted by J_1 and J_2, respectively.

$$
\begin{aligned}
&(E^l \Phi^l_{0,J}) + (E^l A^l_0) - (EA_0) - \varepsilon \\
&\leq (E^l \Phi^l_0) + (E^l A^l_0) - (EA_0) \\
&\leq (E\Phi_0) \\
&\leq (E^u \Phi^u_0) + (E^u A^u_0) - (EA_0) \\
&\leq (E^u \Phi^u_{0,J}) + (E^u A^u_0) - (EA_0) + \varepsilon .
\end{aligned}
$$

Thus, the optimal value $(E\Phi_0)$ of the hydro scheduling problem with risk aversion has upper and lower bounds which are computationally accessible by means of an LP solver. Moreover, the bounds can be made arbitrarily tight by refining the discrete barycentric measures and by improving the piecewise linear approximation of the nonlinear utility function.

It is worthwhile to remark that the evaluation of $(E^l \Phi^l_{0,J})$ and $(E^u \Phi^u_{0,J})$ requires solution of a linear stochastic program with random recourse. In fact, notice that the revenue balance equation (6.9) is bilinear in the decision variables and the stochastic parameters. Generally, stochastic recourse matrices can cause serious trouble (see e.g. [10, Sect. 3.1]), but in this special example they are apparently harmless.

Finally, it should be emphasized that the generalizations of the basic decision problem (6.1), which have been discussed in Sects. 6.2 through 6.5, can easily be combined. However, the correction terms should not thoughtlessly be summed up. For instance, a hydro scheduling problem with a logarithmic utility function and lognormal spot prices can be shown – under certain circumstances – to comply with the regularity conditions (B1)–(B5). Thus, no correction terms are needed.

6.6 Numerical Results

In the remainder we present numerical results of the models proposed in Sects. 6.1 through 6.5. Moreover, we will assess the performance of the newly developed bounding techniques by analyzing the gap between the upper and lower bound on the optimal objective function value. Any bounding technique performs well if the gap becomes small after only few refinements. The deployed algorithms were implemented in C++, and CPLEX 6.6. [16] callable library routines were used to solve the arising linear programs. Moreover, all computations were carried out on a 1 GHz Pentium III PC with 512 MB storage.

6.6.1 Model Parameterization

Consider a hydro scheduling problem with 6 decision stages indexed by $t = 0, \ldots, 5$. Each period between two subsequent stages comprises 28 days. Thus, we are facing a planning horizon of approximately half a year.

Table 6.2. System parameters

Param.	Value	Unit	Range
$\overline{x}_{\mathrm{sell},t}$	1.344E+05	MW h	$\forall t \in \mathcal{T}$
$\overline{x}_{\mathrm{buy},t}$	3.360E+04	MW h	$\forall t \in \mathcal{T}$
$\overline{x}_{\mathrm{store},t}$	4.380E+05	MW h	$\forall t \in \mathcal{T}$
$\underline{x}_{\mathrm{store},T}$	3.500E+05	MW h	
$x_{\mathrm{store},-1}$	3.500E+05	MW h	
ϵ_{P}	7.500E−01		

Table 6.2 lists the relevant technical data of the pumped storage power plant under consideration. Since all periods are of the same length, it is reasonable to assume that the limits on the energy exchanged on the spot market are equal in all decision stages. Moreover, it is intuitively appealing that the reservoir capacity $\overline{x}_{\mathrm{store},t}$ is constant over time, and the amount of energy stored in the reservoir will never drop below 0. However, we require the reservoir to be 80% full at the end of the last period. This is rather a regulatory than a technical restriction, and it takes account of the fact that the model has a finite horizon, whereas real operation of the plant has an indefinite horizon. Without such a restriction, too much energy would be sold in the last decision stage. See [48] for a survey of alternative approaches to mitigate this type of distortions, which are referred to as *end effects* in literature.

Table 6.3. Distributional parameters for the reference problem

Param.	Value	Unit	Range
H_t^{o}	2.608E−01		$\forall t \in \mathcal{T}_{-0}$
H_t^{r}	2.608E−01		$\forall t \in \mathcal{T}_{-0}$
μ_t^{o}	1.848E+01	€ MW^{-1}h^{-1}	$\forall t \in \mathcal{T}_{-0}$
μ_t^{r}	3.696E+03	MW h	$\forall t \in \mathcal{T}_{-0}$
σ_t^{o}	6.869E+00	€ MW^{-1}h^{-1}	$\forall t \in \mathcal{T}_{-0}$
σ_t^{r}	6.869E+03	MW h	$\forall t \in \mathcal{T}_{-0}$
ρ_t	−5.000E−01		$\forall t \in \mathcal{T}_{-0}$
η_0	2.500E+01	€ MW^{-1}h^{-1}	
ξ_0	5.000E+03	MW h	
α_{c}	1.353E−01		

Besides the system parameters we need to specify the distributional parameters of the spot price and inflow processes; cf. Table 6.3. Notice that the increments of prices and inflows are negatively correlated. This assumption can easily be justified. In times of increased precipitation reservoir inflows are high, and hydro producers are forced to generate plenty of power in order to avoid spillage. Then, due to the abundant supply of electric power, spot prices will decrease. Furthermore, the confidence level specified in Table 6.3 entails truncation of each noise term at its mean plus or minus two times its standard deviation, i.e. $n_c = 2$ in (6.2).

The lower bound on $x_{\text{store},T}$ in the last stage induces nontrivial lower bounds on $x_{\text{store},t}$ in all other decision stages (for a discussion of induced constraints see Sect. 2.3). These lower bounds on the reservoir content can be defined recursively, i.e.

$$\underline{x}_{\text{store},t-1} := \max\left\{0, \underline{x}_{\text{store},t} - \xi_{\min,t} - \overline{x}_{\text{buy},t} + \delta\right\}, \quad t \in \tau_{-0}.$$

Thereby, $\xi_{\min,t} := \inf\{\xi_t \in \Xi_t\}$, and $\delta > 0$ is a small energy shift which ensures the existence of Slater points in extreme scenarios. Notice that the hydropower producer can meet the prescribed target value $\underline{x}_{\text{store},T}$ with probability 1 if in every stage $t < T$ the reservoir content lies above $\underline{x}_{\text{store},t}$. This might of course involve intensive pumping activity in certain scenarios. As δ drops to 0, non-anticipativity of the constraint multifunction is preserved. However, for any admissible outcome and decision history with $x_{\text{store},t-1} = \underline{x}_{\text{store},t-1}$ and $\xi_t = \xi_{\min,t}$ there is only one single feasible decision vector, which is certainly no Slater point. To avoid this kind of problems, we fix δ at a very small though strictly positive value in our calculations, i.e. δ=1.000E−06 MW h.

Note that the upper bounds on the reservoir content do not lead to induced constraints. This is a consequence of the fact that inflows ξ_t are much smaller than $\overline{x}_{\text{sell},t}$ in each stage and every scenario for the given parameter setting.

6.6.2 Discretization of the Probability Space

In order to apply the barycentric approximation scheme, we have to find a suitable decomposition of the conditional probability P_t for each $t \in \tau_{-0}$, see Sect. 4.4. For simplicity, we first decompose the probability distribution Q_t of the disturbances $(\varepsilon_t^{\text{o}}, \varepsilon_t^{\text{r}})$ as in (4.30). In particular, we represent Q_t as a convex combination of truncated normal distributions with rectangular supports. Thereby, the supports of two different components are required to be essentially disjoint, i.e. their intersection must be a Q_t-null set. Fig. 6.1 shows the supports of all components corresponding to four successively refined partitions of Q_t; diagram (a) represents the trivial partition, whereas the diagrams (b), (c), and (d) visualize partitions with 2, 4, and 5 components, respectively. Note that the grey shaded area depicts the 87.47% confidence ellipse of the underlying normal distribution ($\alpha_c = 13.53\% \iff n_c = 2$). The refinement strategy visualized in Fig. 6.1 was empirically found to perform well; however, other choices are possible.

The decomposition of the probability measure Q_t naturally induces a decomposition of the conditional probability P_t, each of whose components is of the form (4.31). Subsequently, we proceed according to the guidelines of Sect. 4.4. The components of a given P_t-partition are discretized separately

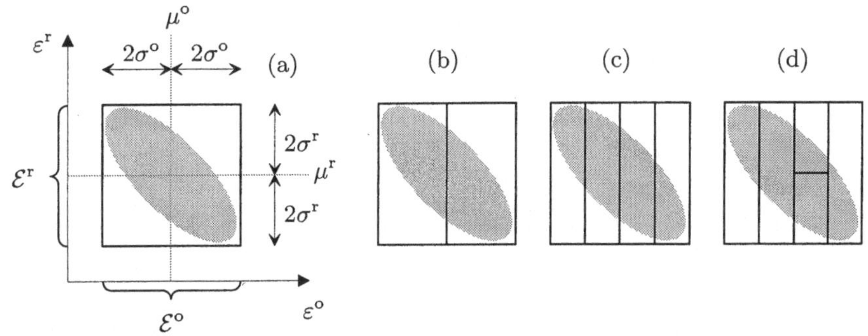

Fig. 6.1. Refinement strategy for the reference problem. Time indices are suppressed for better readability

and then combined to yield the barycentric transition probabilities. Notice that each component contributes two discretization points to the lower and upper barycentric transition probabilities, respectively. Thus, the partitions in (a), (b), (c), and (d) yield scenario trees with branching factors 2, 4, 8, and 10, respectively. The discrete barycentric measures are obtained by composition of the transition probabilities as in (4.36).

Construction of the discrete barycentric probability measures requires evaluation of the first order moments and cross moments of prices and inflows conditional on the outcome history. Since truncated normal distributions are analytically untractable, we use Mathematica 4.1 [72] to calculate these moments by means of standardized numerical integration techniques.

6.6.3 Results of the Reference Problem

In a first step we study the basic hydro scheduling problem without complicating features. Table 6.4 contains the optimal values of the auxiliary stochastic programs associated with the discrete barycentric measures. As argued in Sect. 6.1, the optima of these auxiliary problems represent deterministic lower and upper bounds on the optimum of the original problem. By using the refinement strategy of Fig. 6.1, the relative error between the bounds can be made smaller than 3%. Notice that the optima of the lower (upper) auxiliary stochastic programs are monotonously increasing (decreasing) as the barycentric measures are refined. This observation is consistent with the statements of proposition 4.6.

Convergence of the bounds due to refinements is visualized in Fig. 6.2. Notice that, here, an increase of the branching factor from 8 to 10 improves

Table 6.4. Results of the reference problem

Refinements	Bounds		Error	
Branching factor (–)	$(E^l \Phi_0^l)$ (€)	$(E^u \Phi_0^u)$ (€)	abs. (€)	rel. (–)
2	7.601E+06	1.005E+07	2.450E+06	3.223E−01
4	8.155E+06	8.558E+06	4.025E+05	4.936E−02
8	8.176E+06	8.413E+06	2.376E+05	2.906E−02
10	8.183E+06	8.409E+06	2.270E+05	2.774E−02

accuracy only slightly. Intuitively, one would expect a significant improvement by doubling the branching factor. Then, however, the associated linear programs become too large to be solved.

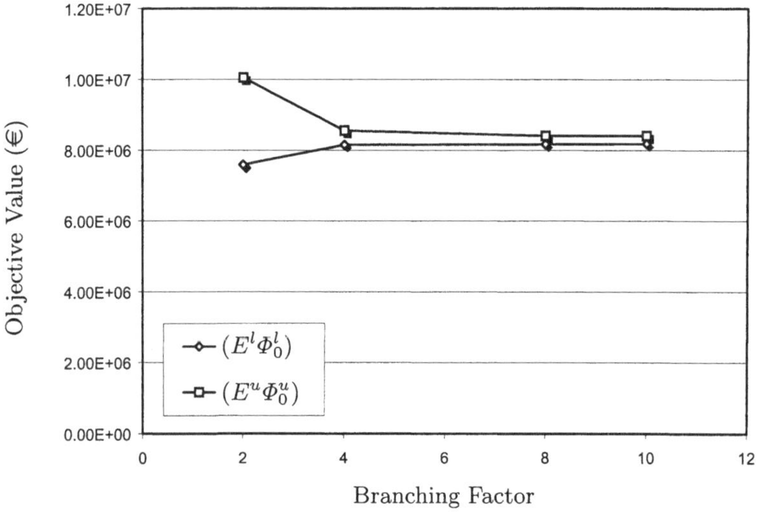

Fig. 6.2. Convergence of bounds due to refinements

6.6.4 Hydro Scheduling Problem with Market Power

Let us now investigate the decision problem with market power, and fix the inverse demand elasticity at c_t :=2.121E−06 MW^{-1}h^{-1} for all $t \in \tau$ (see Sect. 6.2). With this choice, the output sensitive profit functions differ from the inelastic profit functions of Sect. 6.1 by up to 30%, which implies that the hydropower producer has significant market power. In order to solve the decision problem with elastic prices by means of an LP solver, we have to linearize the quadratic functions f_t as in (6.4). Here, we use a piecewise linearization $f_{t,1}$ with 5 segments, which are completely determined by the data in Table 6.5. Note that the parameter labelling this linearization is set to $J = 1$. An elementary calculation shows that f_t and its approximation $f_{t,1}$ differ at most by 330 MW h for any admissible argument $x_{\text{net},t}$. Then, with $T + 1 = 6$ and $E(\eta_t) = 25\,€\,\text{MW}^{-1}\text{h}^{-1}$ for all $t \in \tau$, the tolerance ε in the estimate (6.5) can be set to 5.000E+04 €.

Table 6.5. Piecewise linearization of the inverse demand function

i	$f_{t,1,i}$ (MW h)	$f'_{t,1,i}$ (–)
1	1.071E+00	2.652E+02
2	9.288E–01	2.652E+02
3	7.863E–01	5.053E+03
4	6.438E–01	1.463E+04
5	5.013E–01	2.899E+04

Table 6.6. Results of the hydro scheduling problem with market power

Refinements	**Bounds**		**Error**	
Branching factor (–)	$(E^l\Phi^l_{0,1}) - \varepsilon$ (€)	$(E^u\Phi^u_{0,1}) + \varepsilon$ (€)	abs. (€)	rel. (–)
2	6.574E+06	7.737E+06	1.163E+06	1.770E–01
4	6.708E+06	7.084E+06	3.754E+05	5.595E–02
8	6.749E+06	6.968E+06	2.189E+05	3.244E–02
10	6.757E+06	6.966E+06	2.092E+05	3.096E–02

By convention, $(E^l\Phi^l_{0,1})$ and $(E^u\Phi^u_{0,1})$ denote the optimal values of the hydro scheduling problem with market power if the quadratic functions $\{f_t\}_{t\in\tau}$ are replaced by $\{f_{t,1}\}_{t\in\tau}$, and the original measure P is substituted by the discrete barycentric measures P^l and P^u, respectively. As pointed out in Sect. 6.2, the optimum of the decision problem with market power is squeezed between the optimal values of the auxiliary stochastic programs shifted by the tolerance ε. Table 6.6 contains the lower and upper bounds on the optimal objective value. Moreover, Fig. 6.3 shows convergence of the bounds under the refinement strategy of Fig. 6.1.

6.6.5 Hydro Scheduling Problem with Lognormal Prices

Next, turn to the decision problem of Sect. 6.3. In this model η_t is given a new interpretation as the logarithm of the spot price at time t. Nevertheless, in order to obtain prices of the same order of magnitude as in the reference calculation, the distributional parameters of the random variables $\{\eta_t\}_{t\in\tau}$

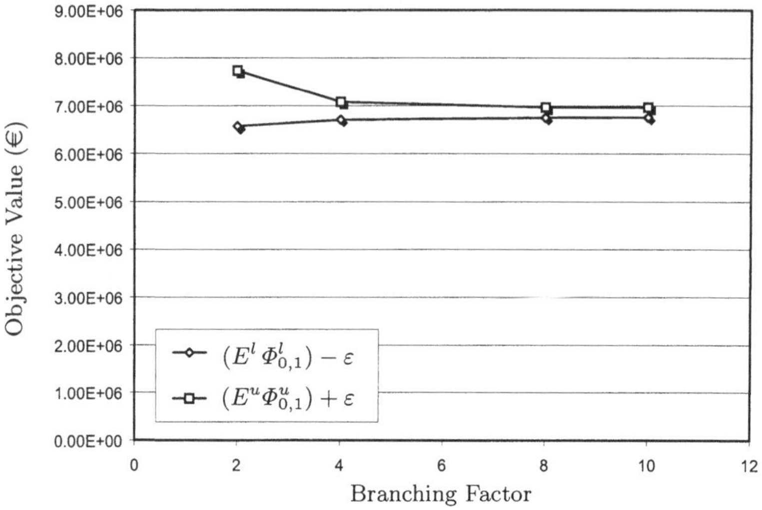

Fig. 6.3. Bounds in the presence of elastic prices

should be suitably modified. The adjusted parameters are listed in Table 6.7. Recalibration of the stochastic processes goes along with the specification of a new refinement strategy; cf. Fig. 6.4. This strategy has empirically been found to be effective, and it almost coincides with the one of Fig. 6.1.

Table 6.7. Distributional parameters in the presence of lognormal prices

Param.	Value	Unit	Range
H^{o}_t	2.608E−01		$\forall t \in \mathcal{T}_{-0}$
H^{r}_t	2.608E−01		$\forall t \in \mathcal{T}_{-0}$
μ^{o}_t	2.048E+00		$\forall t \in \mathcal{T}_{-0}$
μ^{r}_t	3.696E+03	MW h	$\forall t \in \mathcal{T}_{-0}$
σ^{o}_t	4.274E−01		$\forall t \in \mathcal{T}_{-0}$
σ^{r}_t	6.869E+03	MW h	$\forall t \in \mathcal{T}_{-0}$
ρ_t	−5.000E−01		$\forall t \in \mathcal{T}_{-0}$
η_0	3.219E+00		
ξ_0	5.000E+03	MW h	
α_{c}	1.353E−01		

As argued in Sect. 6.3, the determination of bounds on the optimal objective function value involves solution of the auxiliary stochastic programs (4.40) as well as calculation of the expectation values (EA_0), $(E^l A^l_0)$, and $(E^u A^u_0)$. The auxiliary stochastic programs can be tackled in the usual way by means of an LP solver, and their optimal values are listed in the upper panel of Table 6.8 (recall that we work with the refinement strategy of Fig. 6.4). Since the

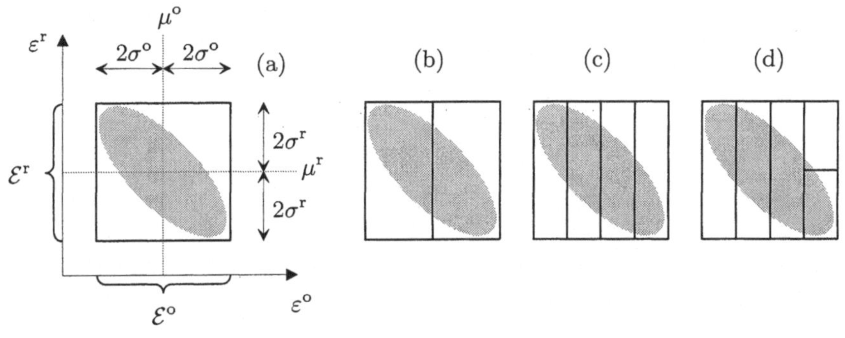

Fig. 6.4. Refinement strategy for the hydro scheduling problem with lognormal prices. Time indices are suppressed for better readability

barycentric measures P^l and P^u are discrete, evaluation of $(E^l A_0^l)$ and $(E^u A_0^u)$ reduces to the calculation of two finite sums; the numerical values in the middle panel of Table 6.8 are based on the choice $X_{\text{net},t}^- :=$3.360E+04 MW h for all $t \in \tau$. In contrast, evaluation of (EA_0) requires integration of the correction terms with respect to the original probability measure. By using serial independence of the disturbances $\{\varepsilon_t^{\rm o}\}_{t\in\tau_{-0}}$, we may write

$$(EA_0) = \sum_{t=1}^{T} X_{\text{net},t}^- \exp(H_{1,t}^{\rm o}\, \eta_0) \prod_{s=1}^{t} E(\exp(H_{s+1,t}^{\rm o}\, \varepsilon_s^{\rm o})).$$

Then, expectation of the exponentiated disturbances under the truncated normal distribution is calculated numerically with the aid of Mathematica. For the given parameter values we obtain $(EA_0) =$2.977E+06 €.

The bounds on the optimal value of the original problem can now be calculated by suitably combining the optimal values of the auxiliary stochastic programs and the expectation values (EA_0), $(E^l A_0^l)$, and $(E^u A_0^u)$; cf. the lower panel of Table 6.8. Figure 6.5 visualizes the development of the bounds under the refinement strategy of Fig. 6.4. Observe that the bounds are still close to the optimal values of the auxiliary stochastic programs.

6.6.6 Hydro Scheduling Problem with Lognormal Inflows

This paragraph is devoted to the study of the decision problem with lognormal inflows, which has been described in Sect. 6.4. Thus, ξ_t no longer represents the reservoir inflow in period t, but its logarithm. To obtain inflows of the same order of magnitude as in the reference problem, we have to recalibrate the stochastic process $\{\xi_t\}_{t\in\tau}$. Table 6.9 contains the adjusted distributional

Table 6.8. Results of the hydro scheduling problem with lognormal prices

Refinements	**Optima of auxiliary stochastic programs**		**Error**	
Branching factor (–)	$(E^l \Phi_0^l)$ (€)	$(E^u \Phi_0^u)$ (€)	abs. (€)	rel. (–)
2	6.228E+06	1.020E+07	3.969E+06	6.372E−01
4	6.801E+06	7.937E+06	1.136E+06	1.671E−01
8	6.927E+06	7.385E+06	4.580E+05	6.612E−02
10	6.934E+06	7.360E+06	4.264E+05	6.150E−02
Refinements	**Expected correction terms**		**Error**	
Branching factor (–)	$(E^l A_0^l)$ (€)	$(E^u A_0^u)$ (€)	abs. (€)	rel. (–)
2	2.776E+06	3.924E+06	1.149E+06	4.139E−01
4	2.885E+06	3.176E+06	2.903E+05	1.006E−01
8	2.925E+06	3.025E+06	1.004E+05	3.432E−02
10	2.928E+06	3.017E+06	8.921E+04	3.047E−02
Refinements	**Bounds**		**Error**	
Branching factor (–)	$(E^l \Phi_0^l) + (E^l A_0^l) - (EA_0)$ (€)	$(E^u \Phi_0^u) + (E^u A_0^u) - (EA_0)$ (€)	abs. (€)	rel. (–)
2	6.027E+06	1.114E+07	5.118E+06	8.491E−01
4	6.709E+06	8.136E+06	1.427E+06	2.127E−01
8	6.875E+06	7.434E+06	5.584E+05	8.122E−02
10	6.884E+06	7.400E+06	5.156E+05	7.489E−02

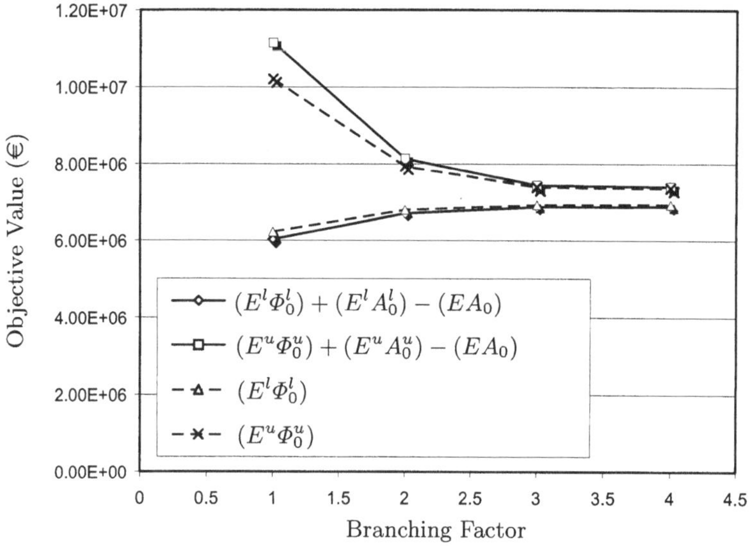

Fig. 6.5. Adjusted bounds in the presence of lognormal prices

parameters, which will be used henceforth. Note that, here again, η_t stands for the spot price of energy at stage t, and not for its logarithm.

Table 6.9. Distributional parameters in the presence of lognormal inflows

Param.	Value	Unit	Range
H_t^{o}	2.608E−01		$\forall t \in \mathcal{T}_{-0}$
H_t^{r}	2.608E−01		$\forall t \in \mathcal{T}_{-0}$
μ_t^{o}	1.848E+01	€MW^{-1}h^{-1}	$\forall t \in \mathcal{T}_{-0}$
μ_t^{r}	7.934E+00		$\forall t \in \mathcal{T}_{-0}$
σ_t^{o}	6.869E+00	€MW^{-1}h^{-1}	$\forall t \in \mathcal{T}_{-0}$
σ_t^{r}	1.566E−01		$\forall t \in \mathcal{T}_{-0}$
ρ_t	−5.000E−01		$\forall t \in \mathcal{T}_{-0}$
η_0	2.500E+01	€MW^{-1}h^{-1}	
ξ_0	1.082E+01		
α_{c}	1.353E−01		

We achieve good convergence behavior of the bounds on the objective value by using the refinement strategy of Fig. 6.6. Since the involved correction terms only depend on the logarithmic inflows, subdivision of the simplex $\mathcal{E}^{\mathrm{r}}$ is necessary for convergence of $(E^l A_0^l)$ and $(E^u A_0^u)$ to the expectation value (EA_0) (see e.g. Fig. 6.6 (c)).

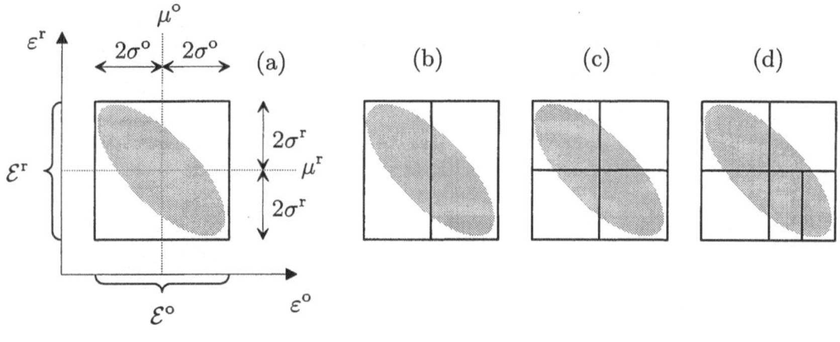

Fig. 6.6. Refinement strategy for the hydro scheduling problem with lognormal inflows. Time indices are suppressed for better readability

As usual, a powerful LP solver can cope with the auxiliary stochastic programs; the corresponding objective values are listed in the upper panel of Table 6.10. Furthermore, the expectation values $(E^l A_0^l)$ and $(E^u A_0^u)$ are easily evaluated as a weighted sum over the scenarios of the discrete measures P^l and P^u, respectively. The numerical values in the middle panel of Table 6.10 are based on the choice $D^{*+}_{\mathrm{bal},t}$:=5.867E+01 € $\mathrm{MW}^{-1}\mathrm{h}^{-1}$, $t \in \tau$. In fact, for the given confidence level α_c =1.353E−01, the constant $D^{*+}_{\mathrm{bal},t}$ represents a strict uniform upper bound on the shadow price of energy at stage t (note that the spot price is always smaller than 44 € $\mathrm{MW}^{-1}\mathrm{h}^{-1}$; dividing this value by the energy conversion factor ϵ_{P} yields the proposed constant $D^{*+}_{\mathrm{bal},t}$). Evaluation of (EA_0) is more complicated as it requires integration of the correction terms with respect to the absolutely continuous probability measure P. By using serial independence of the disturbances $\{\varepsilon^{\mathrm{r}}_t\}_{t\in\tau_{-0}}$, we obtain

$$(EA_0) = -\sum_{t=1}^{T} D^{*+}_{\mathrm{bal},t} \exp(H^{\mathrm{r}}_{1,t}\,\xi_0) \prod_{s=1}^{t} E(\exp(H^{\mathrm{r}}_{s+1,t}\,\varepsilon^{\mathrm{r}}_s)).$$

As before, expectation of the exponentiated disturbances under the truncated normal distribution is calculated numerically with the aid of Mathematica. For the given parameter setting we obtain $(EA_0) = -1.366\mathrm{E}{+}07$ €.

Table 6.10. Results of the hydro scheduling problem with lognormal inflows

Refinements	**Optima of auxiliary stochastic programs**		**Error**	
Branching factor (–)	$(E^l \Phi^l_0)$ (€)	$(E^u \Phi^u_0)$ (€)	abs. (€)	rel. (–)
2	7.401E+06	1.020E+07	2.797E+06	3.779E−01
4	7.952E+06	8.062E+06	1.108E+05	1.393E−02
8	7.784E+06	8.080E+06	2.968E+05	3.813E−02
10	7.787E+06	8.026E+06	2.391E+05	3.070E−02
Refinements	**Expected correction terms**		**Error**	
Branching factor (–)	$(E^l A^l_0)$ (€)	$(E^u A^u_0)$ (€)	abs. (€)	rel. (–)
2	−1.423E+07	−1.353E+07	7.048E+05	4.951E−02
4	−1.423E+07	−1.354E+07	6.935E+05	4.872E−02
8	−1.378E+07	−1.360E+07	1.753E+05	1.272E−02
10	−1.378E+07	−1.360E+07	1.751E+05	1.271E−02
Refinements	**Bounds**		**Error**	
Branching factor (–)	$(E^l \Phi^l_0) + (E^l A^l_0) - (EA_0)$ (€)	$(E^u \Phi^u_0) + (E^u A^u_0) - (EA_0)$ (€)	abs. (€)	rel. (–)
2	6.825E+06	1.033E+07	3.501E+06	5.131E−01
4	7.376E+06	8.180E+06	8.042E+05	1.090E−01
8	7.664E+06	8.136E+06	4.721E+05	6.160E−02
10	7.667E+06	8.081E+06	4.142E+05	5.402E−02

The bounds on the true objective value, which can be calculated in the usual way, are listed in the lower panel of Table 6.10. Moreover, these bounds as well as the optimal values of the auxiliary stochastic programs are visualized in Fig. 6.7.

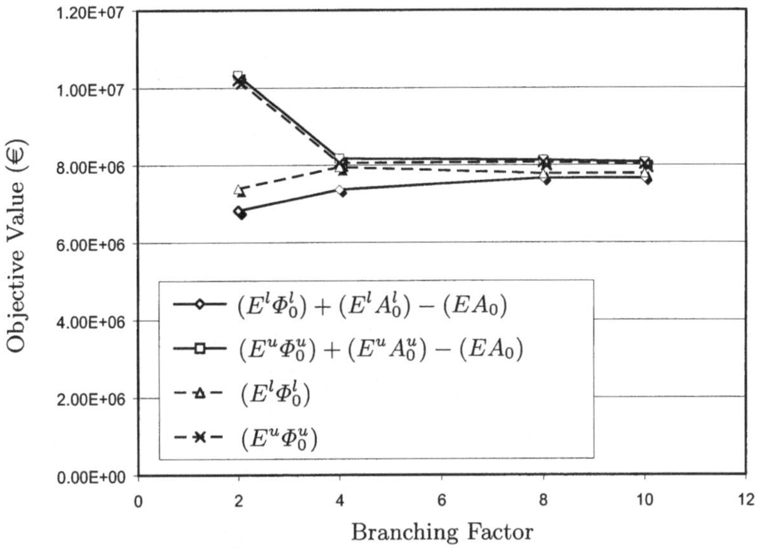

Fig. 6.7. Adjusted bounds in the presence of lognormal inflows

Observe that the gap between the bounds decreases monotonously as the branching factor is swept. However, the difference between $(E^l\Phi_0^l)$ and $(E^u\Phi_0^u)$ is oscillating. This is clear evidence that the saddle structure of at least some recourse functions $\{\Phi_t^l\}_{t\in\mathcal{T}}$ or $\{\Phi_t^u\}_{t\in\mathcal{T}}$ is really lost in the present example. Otherwise, monotonicity should hold due to proposition 4.6.

6.6.7 Hydro Scheduling Problem with Risk-Aversion

Finally, let us study the decision problem with risk-aversion. For simplicity, we work with the parameter setting of the reference problem. Thus, $\{\eta_t\}_{t\in\mathcal{T}}$ and $\{\xi_t\}_{t\in\mathcal{T}}$ are interpreted as the spot price and the inflow process, respectively, with distributional parameters given by Table 6.3. Discretization of the probability space is based on the refinement strategy of Fig. 6.1. Moreover, let us assume that the decision maker's attitude towards risk is characterized by a concave quadratic utility function of the form

$$U(x_{\mathrm{rev},T}) := U_{\mathrm{l}}\, x_{\mathrm{rev},T} - U_{\mathrm{q}}\, x_{\mathrm{rev},T}^2 \quad \text{with} \quad \begin{cases} U_{\mathrm{l}} := 1.700\mathrm{E}{+}00\,€^{-1}, \\ U_{\mathrm{q}} := 5.000\mathrm{E}{-}08\,€^{-2}. \end{cases}$$

By definition, U assigns a dimensionless utility index to every amount $x_{\mathrm{rev},T}$ of wealth accumulated over the entire planning horizon (see also Sect. 6.5). Since the spot price of energy is smaller than 44 €/MWh at any time, rough upper and lower estimates for total wealth $x_{\mathrm{rev},T}$ over all outcomes and all decision strategies feasible in (4.40) are given by $\overline{x}_{\text{rev},T} := 3.548\text{E}{+}07$ € and $\underline{x}_{\text{rev},T} := -8.870\text{E}{+}06$ €, respectively.

Table 6.11. Piecewise linearization of the utility function

i	$U_{1,i}$ (–)	$U'_{1,i}$ (€$^{-1}$)	i	$U_{1,i}$ (–)	$U'_{1,i}$ (€$^{-1}$)
1	4.950E+06	2.7	9	9.750E+06	0.3
2	2.400E+06	2.4	10	1.440E+07	0.0
3	7.500E+05	2.1	11	1.995E+07	−0.3
4	0.000E+00	1.8	12	2.640E+07	−0.6
5	1.500E+05	1.5	13	3.375E+07	−0.9
6	1.200E+06	1.2	14	4.200E+07	−1.2
7	3.150E+06	0.9	15	5.115E+07	−1.5
8	6.000E+06	0.6	16	6.120E+07	−1.8

In order to tackle the hydro scheduling problem with risk-aversion by means of an LP solver, we linearize the quadratic utility function U as in (6.10). Here, we choose a piecewise linearization U_1 with 16 segments, each of which is determined by the data in Table 6.11. Notice that the parameter labelling this linearization is set to $J = 1$. A tedious though elementary calculation shows that U_1 differs from U at most by the tolerance ε :=5.000E+04 on the compact interval $[\underline{x}_{\text{rev},T}, \overline{x}_{\text{rev},T}]$.

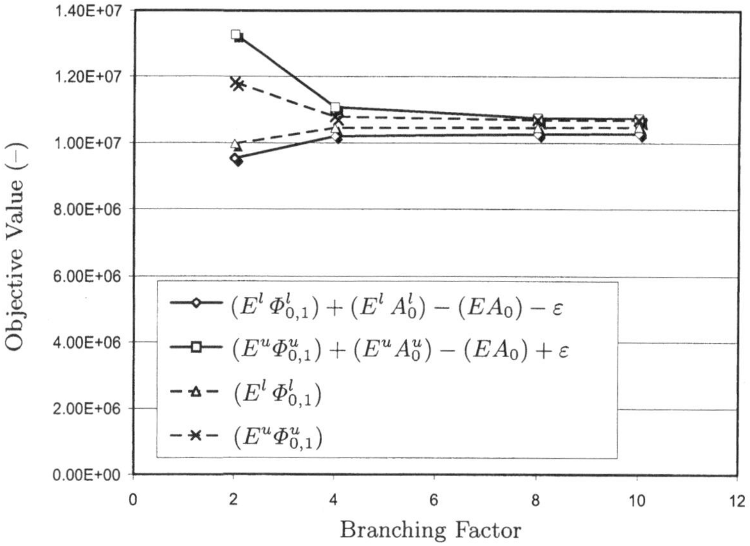

Fig. 6.8. Adjusted bounds in the presence of risk aversion

Table 6.12. Results of the hydro scheduling problem with risk aversion

Refinements	**Optima of auxiliary stochastic programs**		**Error**	
Branching factor (–)	$(E^l\Phi^l_{0,1})$ (–)	$(E^u\Phi^u_{0,1})$ (–)	abs. (–)	rel. (–)
2	9.979E+06	1.182E+07	1.838E+06	1.842E−01
4	1.046E+07	1.079E+07	3.342E+05	3.196E−02
8	1.046E+07	1.069E+07	2.297E+05	2.196E−02
10	1.048E+07	1.069E+07	2.067E+05	1.972E−02
Refinements	**Expected correction terms**		**Error**	
Branching factor (–)	$(E^lA^l_0)$ (–)	$(E^uA^u_0)$ (–)	abs. (–)	rel. (–)
2	6.787E+06	8.579E+06	1.792E+06	2.640E−01
4	6.969E+06	7.418E+06	4.486E+05	6.436E−02
8	7.032E+06	7.191E+06	1.585E+05	2.254E−02
10	7.033E+06	7.191E+06	1.583E+05	2.252E−02
Refinements	**Bounds**		**Error**	
Branching factor (–)	$(E^l\Phi^l_{0,1})+(E^lA^l_0)-(EA_0)-\varepsilon$ (–)	$(E^u\Phi^u_{0,1})+(E^uA^u_0)-(EA_0)+\varepsilon$ (–)	abs. (–)	rel. (–)
2	9.537E+06	1.327E+07	3.730E+06	3.911E−01
4	1.020E+07	1.108E+07	8.828E+05	8.656E−02
8	1.027E+07	1.075E+07	4.883E+05	4.756E−02
10	1.028E+07	1.075E+07	4.651E+05	4.522E−02

Denote by $(E^l\Phi^l_{0,1})$ and $(E^u\Phi^u_{0,1})$ the optimal values of the hydro scheduling problem with risk-aversion if the quadratic utility function U is replaced by U_1, and the original measure P is substituted by the discrete barycentric measures P^l and P^u, respectively. These optimal values can be calculated with a standard LP solver; cf. the upper panel of Table 6.12. As pointed out in Sect. 6.5, the optimum of the decision problem with risk-aversion is squeezed between the optimal values of the auxiliary stochastic programs shifted by a combination of the random variables $(E^lA^l_0)$, $(E^uA^u_0)$, (EA_0), and the tolerance ε. As usual, evaluation of $(E^lA^l_0)$ and $(E^uA^u_0)$ reduces to the calculation of two finite sums. The numerical values in the middle panel of Table 6.12 correspond to the correction terms proposed in Sect. 6.5 with c_T^2 :=1.084E+04 MW h^2 €$^{-2}$. In contrast, evaluation of (EA_0) is fairly involved as it requires integration of the correction terms with respect to the original probability measure P. Reexpressing the spot prices in terms of the disturbances $\{\varepsilon^o_t\}_{t\in\mathcal{T}_{-0}}$, we obtain

$$(EA_0) = c_T^2 \sum_{t=0}^{T} E\left(\left(H^o_{1,t}\eta_0 + \sum_{s=1}^{t} H^o_{s+1,t}\varepsilon^o_s\right)^2\right).$$

Thus, the expectation value (EA_0) can be written as a linear combination of the first and second order moments of the noise terms, which are conveniently calculated by means of Mathematica. Under the current parameterization we end up with $(EA_0) = 7.180\text{E}{+}06$. Then, the bounds on the true objective value can be calculated according to the recipe of Sect. 6.5, and the corresponding results are listed in the lower panel of Table 6.12. Moreover, these bounds as well as the optimal values of the auxiliary stochastic programs are visualized in Fig. 6.8.

As in the other examples of this section, a suitable refinement strategy can reduce the relative gap between the bounds approximately by a factor 10.

7

Conclusions

7.1 Summary of Main Results

In Chap. 2 we study general multistage stochastic programs with non-anticipative constraint multifunctions. Some important terminology is introduced, and several basic results in the field of stochastic optimization are reviewed. We extend the standard terminology by defining the 'natural domains' $\{Y^t\}_{t\in\mathcal{T}}$ and $\{Z^t\}_{t\in\mathcal{T}}$ as specific coordinate projections of the graph of the constraint multifunction. This definition proves useful in many contexts and facilitates the formulation of precise statements. In addition, we establish the fundamental regularity conditions (A1)–(A5), which ensure equivalence of the static and dynamic versions of a multistage stochastic program.

Chapter 3 addresses convex maximization problems with profit functions being concave and constraint functions being convex in the decision variables for all fixed values of the random parameters. As a principal objective, we study the implications of some new regularity conditions (B1)–(B5), which can be shown to imply (A1)–(A5). If a stochastic program complies with these restrictive conditions, then its recourse functions are subdifferentiable and exhibit a characteristic saddle structure on a convex neighborhood of their natural domains. Our approach relaxes some standard assumptions, which are widely used in stochastic programming literature:

- we require the profit and constraint functions to be saddle-shaped and convex, respectively, on a neighborhood of their natural domains (instead of the entire underlying spaces);
- we mitigate the usual strict feasibility condition by allowing for linear affine equality constraints;
- we require compactness of the natural domains $\{Y^t\}_{t\in\mathcal{T}}$ instead of the level sets $\{\mathrm{lev}_{\leq \mathbf{0}}\, \boldsymbol{f}_t\}_{t\in\mathcal{T}}$, which may even be unbounded here.

Restricting the convexity properties of the profit and constraint functions to a neighborhood of their natural domains has decisive technical advantages (see Chap. 5). In addition, the possibility to incorporate equality constraints into an optimization model is of vital importance for most real-life applications (see e.g. Chap. 6).

Chapter 4 develops the barycentric approximation scheme as a classical bounding method for scenario generation. We recover the well-known result that the optima of the auxiliary stochastic programs associated with the lower and upper barycentric measures provide lower and upper bounds on the true optimum, respectively. These bounds can be made arbitrarily tight by means of suitable partitioning techniques. Here, we propose a partitioning scheme which differs from the standard approach in literature. In fact, we suggest partitioning of the conditional probability measures, instead of their supports. This procedure provides more flexibility for synthesizing scenario trees and circumvents the difficulty of handling 'half open' cross-simplices. Finally, we propose a new method for bounding the optimal policy. In particular, we construct a sequence of compact sets, each of which covers the optimal first stage decisions of the original stochastic program. This sequence is shown to converge to the true optimizer set. Our approach sharpens the classical epi-convergence results, as it provides deterministic bounding sets for the optimal decisions.

Chapter 5 is devoted to the study of convex stochastic programs with a generalized nonconvex dependence on the random parameters. If the profit and constraint functions of some stochastic program are nonconvex in the random parameters, then the corresponding recourse functions generally fail to exhibit a saddle structure. Consequently, the barycentric approximation scheme can not be shown to yield bounds on the optimal value and the recourse functions. We prove that, under certain conditions, the saddle structure can be restored by adding specific random variables to the profit functions. These random variables are referred to as 'correction terms'. Loosely speaking, the correction terms compensate the nonconvexites inherent in the underlying stochastic program, and they are required to be nonanticipative and independent of the decision variables. If such correction terms exist, then we may also infer the existence of bounds on the optimal value and the recourse functions. However, in contrast to the well-behaved problems studied in Chap. 4, the plain stage t recourse functions of the auxiliary stochastic programs no longer provide bounds on the stage t recourse function of the original stochastic program. Instead, a lower (upper) bound is given by the lower (upper) auxiliary recourse function shifted by the conditional expectation of the sum of all correction terms from stages t through T. Thereby, conditional expectation is taken with respect to the difference of the lower (upper) barycentric measure and the original measure. The bounds differ little from the plain auxiliary recourse functions if the correction terms are close to bilinear functionals with respect to the topology of uniform convergence or if the barycentric measures are close to the original probability measure with respect to the weak topology.

We formulate suitable regularity conditions which guarantee the existence of correction terms and, in spite of potential nonconvexities,[1] the existence of bounds. If the nonconvexities occur in the objective function, then the underlying stochastic program is required to satisfy some generalized regularity conditions (C1)–(C5). While implied by (B1)–(B5), these new conditions allow for 'regularizable' profit functions, which need not be convex in the random parameters. Loosely speaking, a profit function is called regularizable if it can be transformed to a continuous saddle function on a convex neighborhood of its natural domain by adding a specific correction term. However, notice that the transformed profit function will generally fail to exhibit a saddle structure on the entire underlying space. It is also shown that the class of regularizable profit functions contains a large class of smooth profit functions.

If the nonconvexities occur in the constraints, then we work with another set of generalized regularity conditions (D1)–(D5), which are implied by (B1)–(B5), as well. These new conditions allow for 'regularizable' constraint functions, which need not be convex in the random parameters. Informally speaking, a constraint function is called regularizable if it can be transformed to a continuous convex function on a convex neighborhood of its natural domain by adding a nonanticipative random vector independent of the decision variables. Such random vectors as well as specific bounding vectors for the dual solutions (i.e. the Lagrange multipliers) enter the definition of suitable correction terms which – as pointed out before – reestablish the saddle structure of the recourse functions. It is shown that these correction terms always exist and that the class of regularizable constraint functions contains a large class of smooth constraint functions.

If the nonconvexities occur both in the objective and the constraints, then we invoke a set of regularity conditions (E1)–(E5), which allow for both regularizable profit and constraint functions. Under these generalized regularity conditions, bounds on the optimal value and the recourse functions can systematically be generated. Furthermore, the bounds can be tightened by means of the partitioning techniques described in Chap. 4.

The full symmetry of the proposed bounding method is revealed when studying *linear* stochastic programs. Then, the profit and constraint functions are regularizable if and only if the objective function coefficients and the right hand side vectors are d.c. (i.e. representable as a difference of convex functions) while the recourse and technology matrices are deterministic. These requirements give rise to the regularity conditions (F1)–(F5). Moreover, the

[1]When talking about 'nonconvexities', here, we always refer to the nonconvex dependence of some mapping on the random parameters (and not on the decision variables).

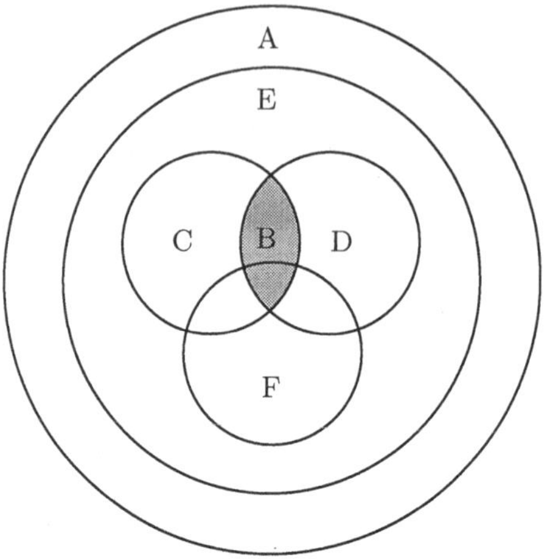

Fig. 7.1. Interdependence of regularity conditions: the set A comprises all stochastic programs subject to the regularity conditions (A1)–(A5), etc. Thus, all problems in A are well-defined and solvable. Moreover, the classical barycentric approximation scheme is applicable for all problems in B, whereas the stochastic programs in C and D may exhibit certain nonconvexities in the objective and the constraints, respectively. E is the largest problem class for which the present work suggests tight bounds. Note that the set F only contains linear stochastic programs

correction terms reflect the intrinsic primal-dual symmetry of linear (stochastic) programs. In fact, the correction terms associated with the nonconvexities in the objective pair specific bounding vectors for the primal solutions with the d.c. components of the objective function coefficients. Conversely, the correction terms associated with the nonconvexities in the constraints pair specific bounding vectors for the dual solutions (i.e. the Lagrange multipliers) with the d.c. components of the right hand side vectors.

In summary, Chap. 5 provides a useful recipe of how the saddle structure of the recourse functions can be reestablished in the presence of certain nonconvexities: loosely speaking, it suffices to add suitable correction terms to the profit functions. After having found such correction terms, any bounding method for scenario generation (that requires the recourse functions to be saddle-shaped) can be used to calculate bounds. Although we usually work with the barycentric approximation scheme, other scenario generation techniques could principally be used (e.g. the method by Edirisinghe [31]). If the random parameters appear only in the objective or in the constraints of a stochastic program, then one might possibly find correction terms which render the recourse functions purely convex or concave, respectively. In this special case, the Jensen and Edmundson-Madansky-type inequalities could be used to derive bounds.

Chapter 6 presents exemplary real-life applications of the theoretical concepts developed in this work. First, we formulate the basic decision problem of a hydropower producer operating a pumped storage power plant. This linear model complies with the regularity conditions (B1)–(B5), and thus the classical barycentric approximation scheme applies, yielding bounds on the optimal value. Next, we study specific generalizations of the basic decision problem. For instance, if the hydro generator has market power, then the objective function becomes quadratic in the decisions. Nevertheless, the model still satisfies the conditions (B1)–(B5), and the barycentric approximation scheme remains applicable. In order to solve the discretized auxiliary stochastic programs by means of an LP solver, the quadratic terms in the objective are approximated by piecewise linear functions. In a next step, we assume that the electricity spot prices are lognormally distributed. This modification may lead to nonconvexities in the objective. However, we argue that the underlying model is linear in the decisions and fulfills the generalized regularity conditions (C1)–(C5). Consequently, application of the barycentric approximation scheme requires that suitable correction terms be added to the profit functions. Subsequently, we consider a model with lognormally distributed reservoir inflows. This model suffers from nonconvexities in the constraints, but it is shown to comply with the generalized regularity conditions (D1)–(D5). Here, as well, the barycentric approximation scheme fails to provide bounds, unless suitable correction terms are introduced. Finally, we investigate a generalized model which maximizes expected utility instead of expected profit. As we work with a concave utility function, which accounts for risk aversion, the objective is nonlinear in the decisions as well as the random parameters. However, the underlying model manifestly satisfies the regularity conditions (C1)–(C5). Thus, the nonconvexities in the random parameters are compensated by adding suitable correction terms, as a consequence of which the use of the barycentric approximation scheme is justified. To tackle the arising auxiliary stochastic programs by means of an LP solver, the concave terms in the objective are approximated by piecewise linear functions.

In any of the above examples, numerical experiments show that the relative gap between the bounds can be reduced to a few percent without exploding the problem size.

7.2 Future Research

In the present work we develop bounds on linear stochastic programs whose objective function coefficients and right hand side vectors are d.c. in the stochastic parameters, while the recourse and technology matrices are deterministic. This result proves satisfactory since any continuous function on a compact domain can be approximated to arbitrary precision by d.c. functions. However, the requirement that the constraint matrices be non-stochastic is restrictive. Future research should be focused on finding ways to relax this condition. It is unsure whether this goal will be accomplished by further improving the bounding measure techniques considered in this work (see Sect. 1.2 for a survey on classical bounding measure techniques). In fact, allowing the constraint matrices to depend on the random parameters can heavily disturb the saddle structure of the recourse functions, and such distortions are usually hard to deal with.

There exist alternative methods which are likely to provide bounds on at least some of the stochastic programs studied in this work. Instead of bounds derived from distributional approximations, one could e.g. work with restricted-recourse bounds, which are proposed in [74] for linear two-stage stochastic programs. One could also invoke scenario tree generation by optimal discretization [85] to derive bounds on stochastic programs with locally Lipschitz recourse functions. Of course, several other approaches might be worth pursuing, as well. The methods proposed in this work as well as possible alternative methods should be assessed as far as their scope, accuracy of the bounds, and computational effort are concerned.

In Sects. 4.6 and 5.5 we suggest bounding sets for the optimal first stage decisions of a given stochastic optimization problem. Calculation of these bounding sets involves the evaluation of the level sets of some concave value function. This, in turn, basically requires solution of a vast number of stochastic programs with slightly varying input parameters. The design and implementation of efficient algorithms to evaluate these compact (and a fortiori bounded) level sets is an important task to be addressed in the future.

A

Conjugate Duality

In this appendix we recall some basic elements of *conjugate duality theory* due to Rockafellar [90] (see also [64, Sect. 2.3]). Important applications of conjugate duality theory in the field of convex stochastic optimization are presented in [92–96]. The methods and terminology introduced here are of vital importance for the derivation of various results in the subsequent appendices and in the main text.

Consider an extended-real-valued function $\hat{\rho} : \mathbb{X} \to [-\infty, \infty]$ on a real topological vector space $\mathbb{X}$. $\hat{\rho}$ is said to be a concave function if its *hypograph*

$$\operatorname{hypo} \hat{\rho} := \{(\boldsymbol{x}, \alpha) | \boldsymbol{x} \in \mathbb{X}, \alpha \in \mathbb{R}, \alpha \leq \hat{\rho}(\boldsymbol{x})\}$$

is a convex subset of the product space $\mathbb{X} \times \mathbb{R}$. The effective domain of a concave function is defined as

$$\operatorname{dom} \hat{\rho} := \{\boldsymbol{x} | \hat{\rho}(\boldsymbol{x}) > -\infty\}.$$

By definition, $\operatorname{dom} \hat{\rho}$ is obtained by projection of $\operatorname{hypo} \hat{\rho}$ on $\mathbb{X}$ and constitutes a convex set. Furthermore, a concave function is called *proper* if $\operatorname{dom} \hat{\rho} \neq \emptyset$ and $\hat{\rho}(\boldsymbol{x}) < \infty$ for all $\boldsymbol{x} \in \mathbb{X}$.

An arbitrary function $\hat{\rho} : \mathbb{X} \to [-\infty, \infty]$ is denoted *upper semicontinuous* (usc) if $\operatorname{hypo} \hat{\rho}$ is a closed set with respect to the product topology on $\mathbb{X} \times \mathbb{R}$. The *usc hull* $\operatorname{usc} \hat{\rho}$ is the smallest usc function $\geq \hat{\rho}$. By construction, we may conclude that

$$\operatorname{hypo} \operatorname{usc} \hat{\rho} = \operatorname{cl} \operatorname{hypo} \hat{\rho}.$$

The concept of usc functions allows us to generalize the closure operation known for sets. We define the *closure* $\operatorname{cl} \hat{\rho}$ of an extended-real-valued function $\hat{\rho}$ through

$$\operatorname{cl} \hat{\rho} := \begin{cases} \operatorname{usc} \hat{\rho} & \text{if} \quad \operatorname{usc} \hat{\rho} < \infty \quad \forall \boldsymbol{x} \in \mathbb{X}, \\ \infty & \text{else.} \end{cases}$$

By convention, $\hat{\rho}$ is called *closed*, if $\hat{\rho} = \operatorname{cl}\hat{\rho}$. Moreover, the *convex hull* $\operatorname{co}\hat{\rho}$ is the smallest concave function $\geq \hat{\rho}$. The hypograph of $\operatorname{co}\hat{\rho}$ consists of the convex hull of the hypograph of $\hat{\rho}$ and a certain set of boundary points (by definition, the 'vertical' fibres of a hypograph must be closed for all fixed vectors $\boldsymbol{x} \in \mathbb{X}$).

$$\operatorname{hypo}\operatorname{co}\hat{\rho} = \{(\boldsymbol{x},\alpha) \in \mathbb{X}\times\mathbb{R} \,|\, (\boldsymbol{x},\beta) \in \operatorname{co}\operatorname{hypo}\hat{\rho} \quad \forall\,\beta < \alpha\}$$

To the (real) linear space $\mathbb{X}$ is associated a dual linear space $\mathbb{X}^*$ along with a bilinear form $\langle\cdot,\cdot\rangle : \mathbb{X}\times\mathbb{X}^* \to \mathbb{R}$. A topology on $\mathbb{X}$ is 'compatible' with this pairing if it is locally convex such that for each $\boldsymbol{x}^* \in \mathbb{X}^*$ the linear functional $\boldsymbol{x} \mapsto \langle\boldsymbol{x}^*,\boldsymbol{x}\rangle$ is continuous, and every continuous linear functional on $\mathbb{X}$ can be represented in this form for some $\boldsymbol{x}^* \in \mathbb{X}^*$. Similarly, a topology on $\mathbb{X}^*$ is compatible with the pairing if it is locally convex such that for each $\boldsymbol{x} \in \mathbb{X}$ the linear functional $\boldsymbol{x}^* \mapsto \langle\boldsymbol{x}^*,\boldsymbol{x}\rangle$ is continuous, and every continuous linear functional on $\mathbb{X}^*$ can be represented in this form for some $\boldsymbol{x} \in \mathbb{X}$. It is assumed that $\mathbb{X}$ and $\mathbb{X}^*$ have been equipped with compatible topologies with respect to the given bilinear form (a topology compatible with a given pairing always exists and can be constructed systematically). Then, the *(concave) conjugate* $\hat{\rho}^*$ of an arbitrary extended-real-valued function $\hat{\rho}$ on X is given by

$$\hat{\rho}^*(\boldsymbol{x}^*) := \inf_{\boldsymbol{x}\in\mathbb{X}} \{\langle\boldsymbol{x}^*,\boldsymbol{x}\rangle - \hat{\rho}(\boldsymbol{x})\}, \tag{A.1}$$

and the *biconjugate* of $\hat{\rho}$ is defined as the conjugate of $\hat{\rho}^*$:

$$\hat{\rho}^{**}(\boldsymbol{x}) := \inf_{\boldsymbol{x}^*\in\mathbb{X}^*} \{\langle\boldsymbol{x}^*,\boldsymbol{x}\rangle - \hat{\rho}^*(\boldsymbol{x}^*)\}. \tag{A.2}$$

Theorem A.1. *For any function $\hat{\rho} : \mathbb{X} \to [-\infty,\infty]$ the following hold:*

(i) $\hat{\rho}^*$ is concave and closed;

(ii) $\hat{\rho}^{**} = \operatorname{cl}\operatorname{co}\hat{\rho}$.

The proof of theorem A.1 relies on a fundamental characterization of closed convex sets, as outlined in the following proposition.

Proposition A.2. *Let C be a subset of a locally convex real vector space $\mathbb{X}$, and assume that C is contained in some closed half-space. Then, the intersection of all the closed half-spaces containing C is $\operatorname{cl}\operatorname{co} C$.*

Proof. Every closed half-space containing C necessarily covers $\operatorname{cl}\operatorname{co} C$. Without loss of generality we may thus assume C to be a strict subset of $\mathbb{X}$ which is convex and closed. Moreover, we may require $\emptyset \neq C$ since the claim is trivial otherwise. It remains to be shown that any such C can be expressed as the intersection of all closed half-spaces that contain it. To this end, choose a vector $\boldsymbol{x} \notin C$. By a standard separation theorem for locally convex spaces (see e.g. theorem 3' in [114, Sect. IV/6]) there exists a closed half-space $H(\boldsymbol{x})$ such that $\boldsymbol{x} \notin H(\boldsymbol{x})$ and $C \subset H(\boldsymbol{x})$. Consequently, the intersection of all closed half-spaces covering C contains no points in the complement of C. □

Proof of theorem A.1. (i) By definition, $\hat{\rho}^*$ is the pointwise infimum of a family of linear affine functions $\boldsymbol{x}^* \mapsto \langle \boldsymbol{x}^*, \boldsymbol{x} \rangle - \hat{\rho}(\boldsymbol{x})$, $\boldsymbol{x} \in \operatorname{dom} \hat{\rho}$. The hypographs of these functions are closed half-spaces in $\mathbb{X} \times \mathbb{R}$; their intersection, i.e. the hypograph of $\hat{\rho}^*$, is a convex closed set. Therefore, the function $\hat{\rho}^*$ is convex and usc. As easily can be seen, $\hat{\rho}^*$ coincides either with the constant function $+\infty$, or $\hat{\rho}^*$ is an usc function which nowhere adopts the value $+\infty$. This notion implies closedness of $\hat{\rho}^*$. Assertion (ii) is proved by reexpressing the biconjugate in terms of $\hat{\rho}$.

$$\hat{\rho}^{**}(\boldsymbol{x}) = \inf_{\boldsymbol{x}^* \in \mathbb{X}^*} \left\{ \langle \boldsymbol{x}^*, \boldsymbol{x} \rangle + \sup_{\hat{\boldsymbol{x}} \in \mathbb{X}} \{ \hat{\rho}(\hat{\boldsymbol{x}}) - \langle \boldsymbol{x}^*, \hat{\boldsymbol{x}} \rangle \} \right\} \tag{A.3}$$

Since the supremum is defined as the least upper bound, we may write

$$\sup_{\hat{\boldsymbol{x}} \in \mathbb{X}} \{ \hat{\rho}(\hat{\boldsymbol{x}}) - \langle \boldsymbol{x}^*, \hat{\boldsymbol{x}} \rangle \} = \inf\{ u \in \mathbb{R} \,|\, u \geq \hat{\rho}(\hat{\boldsymbol{x}}) - \langle \boldsymbol{x}^*, \hat{\boldsymbol{x}} \rangle \quad \forall \hat{\boldsymbol{x}} \in \mathbb{X} \}. \tag{A.4}$$

Substitution of (A.4) into (A.3) shows that $\hat{\rho}^{**}(\boldsymbol{x})$ corresponds to the optimal value of a constrained minimization problem:

$$\begin{aligned} &\inf_{(u,\boldsymbol{x}^*) \in \mathbb{R} \times \mathbb{X}^*} \quad u \\ &\text{s.t.} \quad u + \langle \boldsymbol{x}^*, \hat{\boldsymbol{x}} - \boldsymbol{x} \rangle \geq \hat{\rho}(\hat{\boldsymbol{x}}) \quad \forall \hat{\boldsymbol{x}} \in \mathbb{X}. \end{aligned} \tag{A.5}$$

Apparently, the mathematical program (A.5) has an intuitive geometric interpretation: the feasible set consists of all (continuous) linear affine majorants $U(\hat{\boldsymbol{x}}) = u + \langle \boldsymbol{x}^*, \hat{\boldsymbol{x}} - \boldsymbol{x} \rangle$ of the given function $\hat{\rho}$, and the objective function can be written as $U(\boldsymbol{x})$ (this implies $\hat{\rho}^{**}(\boldsymbol{x}) \geq \hat{\rho}(\boldsymbol{x})$). Equivalently, the hypograph of $\hat{\rho}^{**}$ can be characterized as the intersection of all 'nonvertical' closed half-spaces in $\mathbb{X} \times \mathbb{R}$ covering the hypograph of $\hat{\rho}$ (these nonvertical half-spaces correspond to the hypographs of the linear affine majorants of $\hat{\rho}$). An elementary argument based on proposition A.2 with $C = \operatorname{hypo} \hat{\rho}$ implies $\operatorname{hypo} \hat{\rho}^{**} = \operatorname{cl} \operatorname{co} \operatorname{hypo} \hat{\rho}$. Hence we conclude that $\hat{\rho}^{**} = \operatorname{cl} \operatorname{co} \hat{\rho}$. □

In the sequel we consider an abstract mathematical programming problem over a real topological vector space $\mathbb{X}$.

$$\sup_{\boldsymbol{x} \in \mathbb{X}} \hat{\rho}(\boldsymbol{x}). \tag{A.6}$$

The extended-real-valued objective function $\hat{\rho} : \mathbb{X} \to [-\infty, \infty)$ is assumed to be concave and accounts for possible restrictions on $\boldsymbol{x}$ such that the feasible set of (A.6) is given by $\operatorname{dom} \hat{\rho}$ (see appendix B for an example). The basic idea of duality theory is the embedding of problem (A.6) into a family of parameterized maximization problems. Thereby, the perturbation parameter $\boldsymbol{d}$ ranges over an additional real topological vector space $\mathbb{D}$, and the perturbed objective function $P : \mathbb{X} \times \mathbb{D} \to [-\infty, \infty)$ has the following properties

(a) $P(\boldsymbol{x}, \boldsymbol{0}) = \hat{\rho}(\boldsymbol{x})$ for all $\boldsymbol{x} \in \mathbb{X}$;

(b) $P(\cdot)$ is concave in $(\boldsymbol{x}, \boldsymbol{d})$;

(c) $P(\boldsymbol{x}, \cdot)$ is closed for all $\boldsymbol{x} \in \mathbb{X}$.

We define the optimal value function $\Phi : \mathbb{D} \to [-\infty, \infty]$ as

$$\Phi(\boldsymbol{d}) := \sup_{\boldsymbol{x} \in \mathbb{X}} P(\boldsymbol{x}, \boldsymbol{d}).$$

By construction, the optimal value of the unperturbed problem (A.6) coincides with $\Phi(\boldsymbol{0})$. As outlined in the following proposition, the concavity assumption on P implies that the value function Φ is concave, too.

Proposition A.3. *Φ is a concave function.*

Proof. Denote by E the projection of $\operatorname{hypo} P$ on $\mathbb{D} \times \mathbb{R}$. The hypograph of Φ coincides with E except for some special boundary points (the 'vertical' fibres of $\operatorname{hypo} \Phi$ must be closed for all fixed vectors $\boldsymbol{d} \in \mathbb{D}$):

$$\operatorname{hypo} \Phi = \{(\boldsymbol{d}, \alpha) \in \mathbb{D} \times \mathbb{R} \,|\, (\boldsymbol{d}, \beta) \in E \quad \forall \beta < \alpha\}.$$

By construction, the hypograph of the perturbation function P is convex, and since convexity is preserved under projections, E is convex, as well. This notion entails convexity of $\operatorname{hypo} \Phi$, and therefore Φ is a concave function. □

Let us associate to the linear space $\mathbb{D}$ a dual space $\mathbb{D}^*$ along with a bilinear form $\langle \cdot, \cdot \rangle : \mathbb{D}^* \times \mathbb{D} \to \mathbb{R}$. It is assumed that $\mathbb{D}$ and $\mathbb{D}^*$ have been equipped with compatible topologies with respect to the given bilinear form. Then we are prepared to define the *Lagrangian function*

$$L(\boldsymbol{x}, \boldsymbol{d}^*) := \sup_{\boldsymbol{d} \in \mathbb{D}} \{P(\boldsymbol{x}, \boldsymbol{d}) - \langle \boldsymbol{d}^*, \boldsymbol{d} \rangle\} \equiv -P^*(\boldsymbol{x}, \boldsymbol{d}^*), \tag{A.7}$$

where $P^*(\boldsymbol{x}, \cdot)$ is the conjugate of $P(\boldsymbol{x}, \cdot)$. As usual, such a Lagrangian function can be utilized to define a *primal* maximization problem

$$\sup_{\boldsymbol{x} \in \mathbb{X}} \inf_{\boldsymbol{d}^* \in \mathbb{D}^*} L(\boldsymbol{x}, \boldsymbol{d}^*) \tag{A.8}$$

as well as a corresponding *dual* minimization problem

$$\inf_{\boldsymbol{d}^* \in \mathbb{D}^*} \sup_{\boldsymbol{x} \in \mathbb{X}} L(\boldsymbol{x}, \boldsymbol{d}^*). \tag{A.9}$$

The objective function of the primal problem (A.8) reads

$$\inf_{\boldsymbol{d}^* \in \mathbb{D}^*} L(\boldsymbol{x}, \boldsymbol{d}^*) = \inf_{\boldsymbol{d}^* \in \mathbb{D}^*} \langle \boldsymbol{0}, \boldsymbol{d}^* \rangle - P^*(\boldsymbol{x}, \boldsymbol{d}^*) = P^{**}(\boldsymbol{x}, \boldsymbol{0}) = P(\boldsymbol{x}, \boldsymbol{0}) = \hat{\rho}(\boldsymbol{x}).$$

The third equality follows from proposition A.1, since $P(\boldsymbol{x}, \cdot)$ was assumed to be concave and closed. Therefore, the primal problem (A.8) exactly coincides with the original optimization problem (A.6). For the further study of the mathematical programs (A.8) and (A.9) we need a fundamental result about the interchangeability of 'inf' and 'sup' operators.

Proposition A.4. *Consider an extended-real-valued function f defined on the Cartesian product of two arbitrary sets $\mathbb{X}$ and $\mathbb{Y}$, $f : \mathbb{X} \times \mathbb{Y} \to [-\infty, \infty]$. Then the following hold:*

(i) $\sup_{\boldsymbol{x} \in \mathbb{X}} \sup_{\boldsymbol{y} \in \mathbb{Y}} f(\boldsymbol{x}, \boldsymbol{y}) = \sup_{\boldsymbol{y} \in \mathbb{Y}} \sup_{\boldsymbol{x} \in \mathbb{X}} f(\boldsymbol{x}, \boldsymbol{y})$;

(ii) $\inf_{\boldsymbol{x} \in \mathbb{X}} \inf_{\boldsymbol{y} \in \mathbb{Y}} f(\boldsymbol{x}, \boldsymbol{y}) = \inf_{\boldsymbol{y} \in \mathbb{Y}} \inf_{\boldsymbol{x} \in \mathbb{X}} f(\boldsymbol{x}, \boldsymbol{y})$;

(iii) $\sup_{\boldsymbol{x} \in \mathbb{X}} \inf_{\boldsymbol{y} \in \mathbb{Y}} f(\boldsymbol{x}, \boldsymbol{y}) \le \inf_{\boldsymbol{y} \in \mathbb{Y}} \sup_{\boldsymbol{x} \in \mathbb{X}} f(\boldsymbol{x}, \boldsymbol{y})$;

(iv) $\inf_{\boldsymbol{x} \in \mathbb{X}} \sup_{\boldsymbol{y} \in \mathbb{Y}} f(\boldsymbol{x}, \boldsymbol{y}) \ge \sup_{\boldsymbol{y} \in \mathbb{Y}} \inf_{\boldsymbol{x} \in \mathbb{X}} f(\boldsymbol{x}, \boldsymbol{y})$.

Proof. Assertion (i) basically relies on a sequential application of the involved 'sup' operators:

$$\begin{aligned}
& \sup_{\boldsymbol{y} \in \mathbb{Y}} f(\boldsymbol{x}, \boldsymbol{y}) \ge f(\boldsymbol{x}, \hat{\boldsymbol{y}}) && \forall \hat{\boldsymbol{y}} \in \mathbb{Y},\, \boldsymbol{x} \in \mathbb{X} \\
\Rightarrow\ & \sup_{\boldsymbol{x} \in \mathbb{X}} \sup_{\boldsymbol{y} \in \mathbb{Y}} f(\boldsymbol{x}, \boldsymbol{y}) \ge \sup_{\boldsymbol{x} \in \mathbb{X}} f(\boldsymbol{x}, \hat{\boldsymbol{y}}) && \forall \hat{\boldsymbol{y}} \in \mathbb{Y} \\
\Rightarrow\ & \sup_{\boldsymbol{x} \in \mathbb{X}} \sup_{\boldsymbol{y} \in \mathbb{Y}} f(\boldsymbol{x}, \boldsymbol{y}) \ge \sup_{\boldsymbol{y} \in \mathbb{Y}} \sup_{\boldsymbol{x} \in \mathbb{X}} f(\boldsymbol{x}, \boldsymbol{y}).
\end{aligned}$$

The converse inequality holds by symmetry, and thus the claim is proved. Statement (ii) is equivalent to (i). Moreover, assertion (iii) follows from an analogous argument:

$$\begin{aligned}
& \inf_{\boldsymbol{y} \in \mathbb{Y}} f(\boldsymbol{x}, \boldsymbol{y}) \le f(\boldsymbol{x}, \hat{\boldsymbol{y}}) && \forall \hat{\boldsymbol{y}} \in \mathbb{Y},\, \boldsymbol{x} \in \mathbb{X} \\
\Rightarrow\ & \sup_{\boldsymbol{x} \in \mathbb{X}} \inf_{\boldsymbol{y} \in \mathbb{Y}} f(\boldsymbol{x}, \boldsymbol{y}) \le \sup_{\boldsymbol{x} \in \mathbb{X}} f(\boldsymbol{x}, \hat{\boldsymbol{y}}) && \forall \hat{\boldsymbol{y}} \in \mathbb{Y} \\
\Rightarrow\ & \sup_{\boldsymbol{x} \in \mathbb{X}} \inf_{\boldsymbol{y} \in \mathbb{Y}} f(\boldsymbol{x}, \boldsymbol{y}) \le \inf_{\boldsymbol{y} \in \mathbb{Y}} \sup_{\boldsymbol{x} \in \mathbb{X}} f(\boldsymbol{x}, \boldsymbol{y}).
\end{aligned}$$

Due to lack of symmetry the reversed inequality is not always fulfilled. Finally, statement (iv) is equivalent to (iii). □

Let us now investigate the objective function $\hat{\sigma} : \mathbb{D}^* \to [-\infty, \infty]$ of the dual problem (A.9).

$$
\begin{aligned}
\hat{\sigma}(\boldsymbol{d}^*) &:= \sup_{\boldsymbol{x}\in\mathbb{X}} L(\boldsymbol{x},\boldsymbol{d}^*) \\
&= \sup_{\boldsymbol{x}\in\mathbb{X}} \sup_{\boldsymbol{d}\in\mathbb{D}} \{P(\boldsymbol{x},\boldsymbol{d}) - \langle \boldsymbol{d}^*,\boldsymbol{d}\rangle\} && \text{(definition of } L(\cdot)) \\
&= \sup_{\boldsymbol{d}\in\mathbb{D}} \sup_{\boldsymbol{x}\in\mathbb{X}} \{P(\boldsymbol{x},\boldsymbol{d}) - \langle \boldsymbol{d}^*,\boldsymbol{d}\rangle\} && \text{(proposition A.4)} \qquad \text{(A.10)} \\
&= \sup_{\boldsymbol{d}\in\mathbb{D}} \{\Phi(\boldsymbol{d}) - \langle \boldsymbol{d}^*,\boldsymbol{d}\rangle\} && \text{(definition of } \Phi(\cdot)) \\
&= -\Phi^*(\boldsymbol{d}^*)
\end{aligned}
$$

Thus, the optimal value of the dual problem reduces to

$$\inf_{\boldsymbol{d}^*\in\mathbb{D}^*} \hat{\sigma}(\boldsymbol{d}^*) = \inf_{\boldsymbol{d}^*\in\mathbb{D}^*} \{\langle \boldsymbol{d}^*,\boldsymbol{0}\rangle - \Phi^*(\boldsymbol{d}^*)\} = \Phi^{**}(\boldsymbol{0}).$$

By theorem A.1, the inequality $\Phi \leq \operatorname{cl}\operatorname{co}\Phi = \operatorname{cl}\Phi$ translates to the *weak duality* statement:[1]

$$\sup_{\boldsymbol{x}\in\mathbb{X}} \hat{\rho}(\boldsymbol{x}) \leq \inf_{\boldsymbol{d}^*\in\mathbb{D}^*} \hat{\sigma}(\boldsymbol{d}^*).$$

If $\sup_{\mathbb{X}} \hat{\rho} < \inf_{\mathbb{D}^*} \hat{\sigma}$, the difference $\inf_{D^*} \hat{\sigma} - \sup_{\mathbb{X}} \hat{\rho}$ is referred to as the *duality gap* in literature. Problem (A.6) relative to its embedding in P is called *normal* if $\sup_{\mathbb{X}} \hat{\rho} = \inf_{\mathbb{D}^*} \hat{\sigma}$, i.e. if *strong duality* holds. Moreover, problem (A.6) is called *stable* if it is normal and the supremum in the dual problem is attained, $\sup_{\mathbb{X}} \hat{\rho} = \min_{\mathbb{D}^*} \hat{\sigma}$.

In order to investigate the structural properties of the dual objective function $\hat{\sigma}$, it should be expressed in terms of the value function Φ (cf. (A.10)).

$$\hat{\sigma}(\boldsymbol{d}^*) = \sup_{\boldsymbol{d}\in\mathbb{D}} \{\Phi(\boldsymbol{d}) - \langle \boldsymbol{d}^*,\boldsymbol{d}\rangle\} \tag{A.11}$$

As in the proof of theorem A.1, the 'sup' operator can be eliminated such that (A.11) reduces to a parametric optimization problem over $\mathbb{R}$:

$$\hat{\sigma}(\boldsymbol{d}^*) = \inf_{u\in\mathbb{R}} \{u \,|\, u + \langle \boldsymbol{d}^*,\boldsymbol{d}\rangle \geq \Phi(\boldsymbol{d}) \;\forall \boldsymbol{d}\in\mathbb{D}\}\,. \tag{A.12}$$

Notice that the feasible set of (A.12) is given by a family of parallel linear affine majorants $U(\boldsymbol{d}) = u + \langle \boldsymbol{d}^*,\boldsymbol{d}\rangle$ of the optimal value function Φ (the 'gradient' $\boldsymbol{d}^*$ is fixed whereas the offset u represents a decision variable), and the objective function amounts to $U(\boldsymbol{0})$. From (A.12) it is evident that the choice of the embedding P (and the corresponding value function Φ) substantially influences the dual objective function $\hat{\sigma}$. It is also clear that the dual feasible set $\operatorname{dom}\hat{\sigma}$ is determined by the asymptotic behavior of Φ. Proposition A.5 characterizes the optimal solution set of the dual problem.

[1] Alternatively, from proposition A.4 (iii) it is immediately clear that the supremum of the primal problem (A.8) is smaller or equal to the infimum of the dual problem (A.9).

Proposition A.5. *The following statements are equivalent:*

(i) $\underline{\boldsymbol{d}}^*$ *is an optimal solution of the dual problem (A.9);*

(ii) $\underline{\boldsymbol{d}}^* \in \partial \Phi^{**}(\mathbf{0})$.

Proof. Since $\Phi^{**}(\mathbf{0})$ is the minimal value of the dual problem, it is always true that $\Phi^{**}(\mathbf{0}) \leq \hat{\sigma}(\boldsymbol{d}^*)$ for all $\boldsymbol{d}^* \in \mathbb{D}^*$. Moreover, (A.12) entails the following sequence of equivalent statements:

$$
\begin{aligned}
& \Phi^{**}(\mathbf{0}) = \hat{\sigma}(\underline{\boldsymbol{d}}^*) \\
\iff & \Phi^{**}(\mathbf{0}) + \langle \underline{\boldsymbol{d}}^*, \boldsymbol{d} \rangle \geq \Phi(\boldsymbol{d}) \quad \forall \boldsymbol{d} \in \mathbb{D} \\
\iff & \Phi^{**}(\mathbf{0}) + \langle \underline{\boldsymbol{d}}^*, \boldsymbol{d} \rangle \geq \Phi^{**}(\boldsymbol{d}) \quad \forall \boldsymbol{d} \in \mathbb{D} \\
\iff & \underline{\boldsymbol{d}}^* \in \partial \Phi^{**}(\mathbf{0})
\end{aligned}
$$

Thus, every subgradient $\underline{\boldsymbol{d}}^* \in \partial \Phi^{**}(\mathbf{0})$ constitutes an optimal solution of (A.9) and vice versa. □

Corollary A.6. *The dual problem (A.9) is solvable if and only if* Φ^{**} *is subdifferentiable at the origin.*

B

Lagrangian Duality

Now we exploit the results of appendix A in order to develop a specific *Lagrangian duality* scheme for constrained maximization problems over $\mathbb{X} = \mathbb{R}^n$. In particular, let us study convex mathematical programs of the form

$$\begin{aligned} &\sup\ \rho(\boldsymbol{x}) \\ &\text{s.t.}\quad \boldsymbol{f}(\boldsymbol{x}) \leq \boldsymbol{0}\,. \end{aligned} \tag{B.1}$$

The objective function $\rho : \mathbb{R}^n \to [-\infty, \infty)$ of (B.1) is concave, whereas the vector-valued constraint function $\boldsymbol{f} : \mathbb{R}^n \to \mathbb{R}^r$ is assumed to be (component-wise) convex. As usual, the constraints can be made implicit by passing over to an 'effective' objective function:

$$\hat{\rho}(\boldsymbol{x}) = \begin{cases} \rho(\boldsymbol{x}) \text{ for } \boldsymbol{f}(\boldsymbol{x}) \leq \boldsymbol{0}, \\ -\infty \text{ else.} \end{cases} \tag{B.2}$$

By means of the effective objective function (B.2), the mathematical program (B.1) can be represented as an unconstrained optimization problem (this formulation is compatible with the conventions met in appendix A). For technical reasons we require (B.1) to satisfy some regularity conditions. First, the feasible set $X := \{\boldsymbol{x} | \boldsymbol{f}(\boldsymbol{x}) \leq \boldsymbol{0}\}$ must be a compact subset of $\operatorname{int}\operatorname{dom}\rho$. Furthermore, we assume that ρ is continuous on $\operatorname{int}\operatorname{dom}\rho$, and there is a Slater point $\boldsymbol{y} \in X$ with the property $\boldsymbol{f}(\boldsymbol{y}) < \boldsymbol{0}$. Notice that these specifications do not allow for equality constraints. For didactic reasons, the investigation of equality constraints is postponed to the end of this section.

As outlined in appendix A, the formulation of a dual minimization problem is based on a convex embedding of the primal problem (B.1) into a family of perturbed maximization problems. A convenient embedding is defined through

$$P(\boldsymbol{x}, \boldsymbol{d}) = \begin{cases} \rho(\boldsymbol{x}) \text{ for } \boldsymbol{f}(\boldsymbol{x}) \leq \boldsymbol{d}, \\ -\infty \text{ else,} \end{cases} \tag{B.3}$$

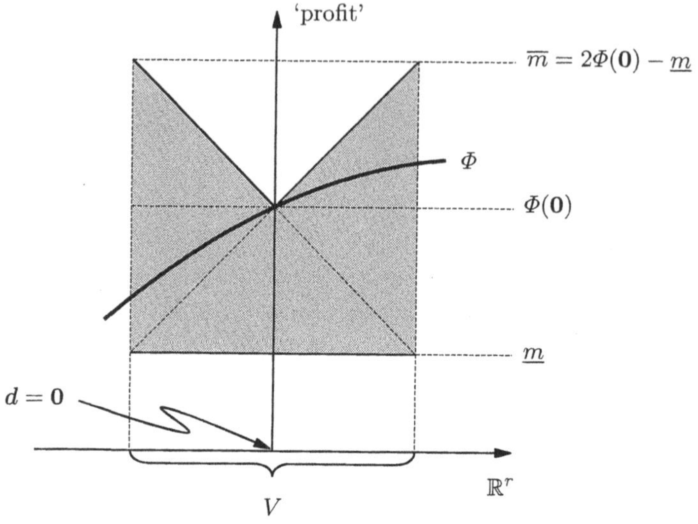

Fig. B.1. The value function Φ is concave and bounded below on a closed ball V centered at the origin. As $\Phi(\mathbf{0})$ is finite, it is geometrically clear that the graph of Φ over V lies within the shaded region and is also bounded above

where the perturbation parameter $\boldsymbol{d}$ ranges over $\mathbb{D} = \mathbb{R}^r$. In many applications, the embedding (B.3) is particularly useful for sensitivity analysis. Some basic properties of the associated value function are summarized in the subsequent proposition.

Proposition B.1. *The value function $\Phi(\boldsymbol{d}) = \sup_{\boldsymbol{x}} P(\boldsymbol{x}, \boldsymbol{d})$ corresponding to the embedding (B.3) has the following properties:*

(i) Φ is proper, concave, and finite on a neighborhood of the origin;

(ii) $\Phi(\boldsymbol{d}) \geq \Phi(\hat{\boldsymbol{d}})$ for $\boldsymbol{d} \geq \hat{\boldsymbol{d}}$, i.e. Φ is monotonous.

Proof. (i) By construction, the embedding P is jointly concave in $(\boldsymbol{x}, \boldsymbol{d})$, and thus proposition A.3 implies that Φ is concave. By the assumptions on the objective function and the constraints, it can easily be verified that $\Phi(\mathbf{0})$ is finite. Moreover, the embedding $P(\boldsymbol{y}, \cdot)$ is continuous on a closed ball V centered at $\boldsymbol{d} = \mathbf{0}$ if the Slater point $\boldsymbol{y}$ is held fixed. Thus, Φ is bounded below by $\underline{m} = \inf\{P(\boldsymbol{y}, \boldsymbol{d}) | \boldsymbol{d} \in V\}$ on V. In addition, concavity and finiteness at the origin entail that Φ is bounded above by $\overline{m} = 2\Phi(\mathbf{0}) - \underline{m}$ on V (cf. Fig. B.1 for the geometric motivation behind this argument). Hence, Φ is finite on V. Again by concavity, we find $\Phi < \infty$ on $\mathbb{R}^r$, which implies properness. (ii) The value function Φ grows with $\boldsymbol{d}$ since the restrictions are relaxed as the perturbation parameter is increased. □

For later use, the effective domain of the value function $D := \operatorname{dom}\Phi$ has to be analyzed.

Corollary B.2. *The assertions of proposition* B.1 *imply:*

(i) D *is non-empty and convex (not necessarily closed);*

(ii) $\boldsymbol{d} \in D \Rightarrow \boldsymbol{d} + \mathbb{R}^r_+ \subset D$ *and* $\boldsymbol{d} \in D^c \Rightarrow \boldsymbol{d} + \mathbb{R}^r_- \subset D^c$.

Since we concentrate on finite-dimensional problems, we do not have to consider paired topological spaces. By means of the inner product, any linear functional $\psi : \mathbb{R}^r \to \mathbb{R}$ on the vector space of perturbation parameters can be identified with an element $\boldsymbol{d}^*$ of $\mathbb{R}^r$, i.e. $\psi(\boldsymbol{d}) = \langle \boldsymbol{d}^*, \boldsymbol{d} \rangle = \sum_{i=1}^r d_i^* \cdot d_i$. With this convention the general definition (A.7) leads to the classical *Lagrangian function*

$$L(\boldsymbol{x}, \boldsymbol{d}^*) = \begin{cases} \rho(\boldsymbol{x}) - \langle \boldsymbol{d}^*, \boldsymbol{f}(\boldsymbol{x}) \rangle & \text{for } \boldsymbol{d}^* \geq \boldsymbol{0}, \\ +\infty & \text{else.} \end{cases} \tag{B.4}$$

The *primal* problem associated with the Lagrangian (B.4) can be written as

$$\sup_{\boldsymbol{x}} \inf_{\boldsymbol{d}^* \geq \boldsymbol{0}} \rho(\boldsymbol{x}) - \langle \boldsymbol{d}^*, \boldsymbol{f}(\boldsymbol{x}) \rangle, \tag{B.5}$$

whereas the *dual* problem reads

$$\inf_{\boldsymbol{d}^* \geq \boldsymbol{0}} \sup_{\boldsymbol{x}} \rho(\boldsymbol{x}) - \langle \boldsymbol{d}^*, \boldsymbol{f}(\boldsymbol{x}) \rangle. \tag{B.6}$$

Apparently, the dual objective function amounts to

$$\hat{\sigma}(\boldsymbol{d}^*) = \begin{cases} \sup_{\boldsymbol{x}} \rho(\boldsymbol{x}) - \langle \boldsymbol{d}^*, \boldsymbol{f}(\boldsymbol{x}) \rangle & \text{for } \boldsymbol{d}^* \geq \boldsymbol{0}, \\ +\infty & \text{else,} \end{cases}$$

and the dual feasible set is determined by $D^* := \operatorname{dom}\hat{\sigma}$.[1] Little surprisingly, D^* is a convex set. However, in contrast to the primal feasible set, D^* is unbounded and not necessarily closed. Below, these results will be discussed in more detail.

Proposition B.3. *The dual feasible set* D^* *is convex and unbounded.*

Proof. Since Φ^* is a concave function, the dual feasible set $D^* = \operatorname{dom}\hat{\sigma} = \operatorname{dom}\Phi^*$ is convex (cf. (A.10)). In order to prove unboundedness of D^*, we have to show in a first step that the effective domain of the value function is not $\mathbb{R}^r$. To this end we introduce the auxiliary function

[1] It is standard practice to define the effective domain of a convex function $\hat{\sigma}$ as $\operatorname{dom}\hat{\sigma} := \{\boldsymbol{d}^* | \hat{\sigma}(\boldsymbol{d}^*) < \infty\}$.

$$\hat{f}(\boldsymbol{x}) := \max\{f_i(\boldsymbol{x}) \,|\, i = 1, \dots r\}.$$

Obviously, $\hat{f}$ is convex and continuous on the entire decision space. Compactness of X implies that $\bar{d} := \min_{\boldsymbol{x} \in X} \hat{f}(\boldsymbol{x})$ is a finite number. Thus we find that

$$\{\boldsymbol{x} \,|\, \boldsymbol{f}(\boldsymbol{x}) \leq d\,\boldsymbol{e}\} = \emptyset \quad \forall d < \bar{d},$$

where $\boldsymbol{e} = (1, \dots, 1)$ is an r-dimensional vector with identical entries. By construction, the vectors $d\,\boldsymbol{e}$ are no elements of D, for all $d < \bar{d}$. Consequently, compactness of X ensures that neither the effective domain of the value function nor its closure coincide with the underlying space of perturbation parameters, i.e. $\operatorname{cl} D \neq \mathbb{R}^r$. Next, choose an arbitrary vector $\boldsymbol{d}_0 \in (\operatorname{cl} D)^{\mathrm{c}}$. According to a standard separation theorem for convex sets, there exists a linear functional $\psi(\boldsymbol{d}) = \langle \boldsymbol{g}^*, \boldsymbol{d} - \boldsymbol{d}_0 \rangle$ such that $\psi(\boldsymbol{d}) > 0$ for all $\boldsymbol{d} \in \operatorname{cl} D$. Note that the vector $\boldsymbol{g}^* \in \mathbb{R}^r$ can be interpreted as the gradient of ψ. The further argumentation relies on the following characterization of the dual feasible set D^*, which is motivated by (A.12).

$$D^* = \{\boldsymbol{d}^* \in \mathbb{R}^r \,|\, \exists\, u \in \mathbb{R} \text{ such that } u + \langle \boldsymbol{d}^*, \boldsymbol{d} \rangle \geq \Phi(\boldsymbol{d}) \;\forall \boldsymbol{d} \in \mathbb{R}^r\} \neq \emptyset$$

Thus, for every $\boldsymbol{d}^* \in D^*$ there exists a real number u and a linear affine function $U(\boldsymbol{d}) = u + \langle \boldsymbol{d}^*, \boldsymbol{d} \rangle$ that majorizes the value function $\Phi(\boldsymbol{d})$. By construction, the linear affine function $U(\boldsymbol{d}) + \lambda\,\psi(\boldsymbol{d})$ is a majorant of $\Phi(\boldsymbol{d})$ as well, for any $\lambda \in \mathbb{R}_+$. Consequently, its gradient $\boldsymbol{d}^* + \lambda\,\boldsymbol{g}^*$ is an element of D^*, and the dual feasible set is unbounded since the parameter λ may tend to infinity. □

The dual feasible set D^* is not only unbounded, but it is not even necessarily closed. For example, consider a one-dimensional convex optimization problem whose objective function is given by

$$\rho(x) := \begin{cases} \ln(x+1) & \text{for } x \geq 0, \\ x & \text{else.} \end{cases}$$

Assume that there is a convex constraint function of the form $f(x) := |x| - 1$. Thus, the feasible set $X = [-1, 1]$ is compact and contains a Slater point. Moreover, ρ is concave, f is convex, and both are continuous. A straightforward calculation shows that the standard embedding (B.3) leads to the following value function

$$\Phi(d) = \sup_x \{\rho(x) \,|\, f(x) \leq d\} = \begin{cases} \ln(d+2) & \text{for } d \geq -1, \\ -\infty & \text{else.} \end{cases}$$

In a next step we determine the conjugate of Φ.

$$\Phi^*(d^*) = \inf_d \{d^*\, d - \Phi(d)\} = \begin{cases} d^* & \text{for } 1 < d^* \\ 1 - 2d^* + \ln(d^*) & \text{for } 0 < d^* \leq 1 \\ -\infty & \text{else} \end{cases}$$

From general theory we know that the dual feasible set D^* is given by the effective domain of Φ^*, and thus $D^* = (0, \infty)$ is an open subset of $\mathbb{R}$.

It is worthwhile to remark that the dual feasible set differs from the set of all subgradients of Φ^{**} only by specific boundary points. Formally speaking, the range of the subdifferential multifunction $\partial\Phi^{**}$ is defined as

$$\operatorname{range}\partial\Phi^{**} := \bigcup_{\boldsymbol{d}\in\mathbb{R}^r} \partial\Phi^{**}(\boldsymbol{d}).$$

Since Φ^{**} is a closed proper concave function, the following identity holds:

$$\operatorname{int} D^* = \operatorname{int}\operatorname{dom}\Phi^* \subset \operatorname{range}\partial\Phi^{**} \subset \operatorname{dom}\Phi^* = D^*$$

(cf. [89, p. 227]). The analysis of the value function and its subdifferential multifunction not only yields a convenient characterization of the dual feasible set but also provides fundamental insights concerning solvability and stability of the primal dual pair of mathematical programs (B.5) and (B.6).

Proposition B.4. *There exists a convex neighborhood V of $\boldsymbol{d} = \boldsymbol{0}$ such that the value function Φ is continuous and subdifferentiable on V.*

Proof. This proposition is closely related to theorem 3.18 and can be proved in a similar manner. By proposition B.1, Φ is proper, concave, and finite on a neighborhood V of the origin. Without loss of generality we may assume that V is open and convex. Thus, by [89, theorem 10.1], Φ is continuous on V. Moreover, [89, theorem 23.4] entails subdifferentiability of Φ on V. □

As Φ is a proper concave function (see proposition B.1), we have $\operatorname{cl}\Phi = \operatorname{usc}\Phi$. In addition, by continuity we find $\operatorname{cl}\Phi = \Phi$ on V. This notion implies $\Phi(\boldsymbol{0}) = \Phi^{**}(\boldsymbol{0})$, and thus the optimization problem (B.1) with the embedding (B.3) is normal (i.e. strong duality holds). Notice that the primal problem (B.5) is solvable since X is compact and the objective function ρ is continuous on X. Moreover, the dual problem (B.6) is solvable as well since $\Phi^{**} = \operatorname{cl}\Phi$ is subdifferentible at the origin (cf. proposition A.6). Hence, the optimization problem under consideration is stable.

If we allow for equality constraints in (B.1), the Lagrangian duality scheme developed above must be slightly modified. Let us therefore investigate a mathematical program of the form

$$\begin{aligned} &\sup\ \rho(\boldsymbol{x}) \\ \text{s.t.}\ \ &\boldsymbol{f}^{\text{in}}(\boldsymbol{x}) \le \boldsymbol{0} \\ &\boldsymbol{f}^{\text{eq}}(\boldsymbol{x}) = \boldsymbol{0}. \end{aligned} \tag{B.7}$$

As before, the objective function $\rho : \mathbb{R}^n \to [-\infty, \infty)$ of (B.7) is concave, whereas the constraint function $\boldsymbol{f}^{\text{in}} : \mathbb{R}^n \to \mathbb{R}^{r^{\text{in}}}$ belonging to the inequality constraints is assumed to be convex. Moreover, the constraint function of the equality constraints $\boldsymbol{f}^{\text{eq}} : \mathbb{R}^n \to \mathbb{R}^{r^{\text{eq}}}$ is required to be linear affine. As before, we impose certain regularity conditions. Above all, the feasible set

$$X := \{\boldsymbol{x} \mid \boldsymbol{f}^{\text{in}}(\boldsymbol{x}) \leq \boldsymbol{0},\ \boldsymbol{f}^{\text{eq}}(\boldsymbol{x}) = \boldsymbol{0}\}$$

must be a compact subset of $\operatorname{int}\operatorname{dom}\rho$. Furthermore, we assume that ρ is continuous on $\operatorname{int}\operatorname{dom}\rho$, and (B.7) satisfies Slater's constraint qualification. Thus, the gradients of the equality constraints are linearly independent, and there is a Slater point $\boldsymbol{y} \in X$ which is feasible and strictly satisfies the inequality constraints, i.e. $\boldsymbol{f}^{\text{in}}(\boldsymbol{y}) < \boldsymbol{0}$ and $\boldsymbol{f}^{\text{eq}}(\boldsymbol{y}) = \boldsymbol{0}$. Then, we may choose an embedding similar to (B.3)

$$P(\boldsymbol{x}, \boldsymbol{d}^{\text{in}}, \boldsymbol{d}^{\text{eq}}) = \begin{cases} \rho(\boldsymbol{x}) \text{ for } \boldsymbol{f}^{\text{in}}(\boldsymbol{x}) \leq \boldsymbol{d}^{\text{in}} \text{ and } \boldsymbol{f}^{\text{eq}}(\boldsymbol{x}) = \boldsymbol{d}^{\text{eq}}, \\ -\infty \text{ else.} \end{cases} \tag{B.8}$$

Apparently, the perturbation parameter $(\boldsymbol{d}^{\text{in}}, \boldsymbol{d}^{\text{eq}})$ ranges over the product space $\mathbb{R}^{r^{\text{in}}} \times \mathbb{R}^{r^{\text{eq}}}$. Some essential properties of the associated optimal value function are provided by the following proposition.

Proposition B.5. *The value function* $\Phi(\boldsymbol{d}^{\text{in}}, \boldsymbol{d}^{\text{eq}}) = \sup_{\boldsymbol{x}} P(\boldsymbol{x}, \boldsymbol{d}^{\text{in}}, \boldsymbol{d}^{\text{eq}})$ *corresponding to the embedding* (B.8) *is proper, concave, and finite on a neighborhood of the origin.*

Proof. By construction, the embedding P is jointly concave in $(\boldsymbol{x}, \boldsymbol{d}^{\text{in}}, \boldsymbol{d}^{\text{eq}})$, and thus proposition A.3 implies that Φ is concave. By the assumptions on the objective function and the constraints, it can easily be verified that $\Phi(\boldsymbol{0}, \boldsymbol{0})$ is finite. Moreover, due to Slater's constraint qualification, there is a closed ball V centered at $(\boldsymbol{d}^{\text{in}}, \boldsymbol{d}^{\text{eq}}) = (\boldsymbol{0}, \boldsymbol{0})$ and a continuous function $\tilde{\boldsymbol{x}} : V \to \mathbb{R}^n$ such that the graph of $\tilde{\boldsymbol{x}}$ is a subset of $\operatorname{int}\operatorname{dom}\rho$,

$$\boldsymbol{f}^{\text{in}}(\tilde{\boldsymbol{x}}(\boldsymbol{d}^{\text{in}}, \boldsymbol{d}^{\text{eq}})) < \boldsymbol{d}^{\text{in}}, \quad \text{and} \quad \boldsymbol{f}^{\text{eq}}(\tilde{\boldsymbol{x}}(\boldsymbol{d}^{\text{in}}, \boldsymbol{d}^{\text{eq}})) = \boldsymbol{d}^{\text{eq}}.$$

This is a direct consequence of proposition 3.2 applied to the parametric maximization problem $\sup_{\boldsymbol{x}} P(\boldsymbol{x}, \boldsymbol{d}^{\text{in}}, \boldsymbol{d}^{\text{eq}})$ at the reference point $(\boldsymbol{d}^{\text{in}}, \boldsymbol{d}^{\text{eq}}) = (\boldsymbol{0}, \boldsymbol{0})$. By what has been said above, $\inf_V \rho \circ \tilde{\boldsymbol{x}}$ is a lower bound for Φ on the closed ball V. In addition, concavity and finiteness at the origin entail that Φ is bounded above on V (use an analog argument as in proposition B.1). Hence, Φ is finite on V. Then, by concavity, we find $\Phi < \infty$ on $\mathbb{R}^{r^{\text{in}}} \times \mathbb{R}^{r^{\text{eq}}}$ which implies properness. □

Little surprisingly, the embedding (B.8) leads to the classical *Lagrangian function*

$$L(\boldsymbol{x}, \boldsymbol{d}^{\text{in}*}, \boldsymbol{d}^{\text{eq}*}) = \begin{cases} \rho(\boldsymbol{x}) - \langle \boldsymbol{d}^{\text{in}*}, \boldsymbol{f}^{\text{in}}(\boldsymbol{x}) \rangle - \langle \boldsymbol{d}^{\text{eq}*}, \boldsymbol{f}^{\text{eq}}(\boldsymbol{x}) \rangle & \boldsymbol{d}^{\text{in}*} \geq \boldsymbol{0}, \\ +\infty & \text{else.} \end{cases} \tag{B.9}$$

The *primal* problem associated with the Lagrangian (B.9) can be written as

$$\sup_{\boldsymbol{x}} \inf_{(\boldsymbol{d}^{\text{in}*}, \boldsymbol{d}^{\text{eq}*})} L(\boldsymbol{x}, \boldsymbol{d}^{\text{in}*}, \boldsymbol{d}^{\text{eq}*}), \tag{B.10}$$

whereas the *dual* problem reads

$$\inf_{(\boldsymbol{d}^{\text{in}*}, \boldsymbol{d}^{\text{eq}*})} \sup_{\boldsymbol{x}} L(\boldsymbol{x}, \boldsymbol{d}^{\text{in}*}, \boldsymbol{d}^{\text{eq}*}). \tag{B.11}$$

As before, it can generally be shown that the dual feasible set is convex and unbounded (see also proposition B.3). By the above reasoning (see proposition B.5), the optimal value function Φ is concave and finite on a convex neighborhood of the origin. Thus, Φ is continuous and subdifferentiable at $(\mathbf{0}, \mathbf{0})$ implying strong duality. Moreover, the primal dual pair (B.10) and (B.11) is stable.

Notice that, by convention, the optimizers of the Lagrangian dual problems (B.6) and (B.11) are usually referred to as *Lagrange multipliers*.

C

Penalty-Based Optimization

In this appendix we develop a penalty-based formulation of the convex maximization problem (B.1). Proposition C.1 and corollary C.2 below are important for the proof of proposition 5.12 in the main text. Here, the regularity conditions of appendix B are still assumed to hold true. Let $\boldsymbol{d}^*_{\text{opt}} \in \mathbb{R}^r$ be some minimizer of the dual problem (B.6) associated with (B.1), and choose an arbitrary bounding vector $\boldsymbol{D}^* \geq \boldsymbol{d}^*_{\text{opt}}$. Then, we may introduce an equivalent unconstrained optimization problem, whose objective is concave on $\mathbb{R}^n$ and continuous on $\operatorname{int}\operatorname{dom}\rho$.

$$\sup_{\boldsymbol{x}} \; \rho(\boldsymbol{x}) - \langle \boldsymbol{D}^*, [\boldsymbol{f}(\boldsymbol{x})]^+ \rangle. \tag{C.1}$$

Notice that the objective function of (C.1) coincides with ρ on the original feasible set X, and any violation of the constraints is penalized proportionally to the bounding vector $\boldsymbol{D}^*$.

Proposition C.1. *The mathematical programs (B.1) and (C.1) have the same optimal values.*

Proof. Strong duality guarantees that

$$\inf_{\boldsymbol{d}^* \geq \boldsymbol{0}} \; \sup_{\boldsymbol{x}} \; \rho(\boldsymbol{x}) - \langle \boldsymbol{d}^*, \boldsymbol{f}(\boldsymbol{x}) \rangle \tag{C.2}$$

has the same optimal value as (B.1). By assumption, $\boldsymbol{D}^*$ represents an upper bounding vector for at least some solutions of (C.2). Therefore, the following optimization problem is equivalent to (C.2).

$$\begin{aligned} \inf_{\boldsymbol{d}^*_t} \; \sup_{\boldsymbol{x}} \; & \rho(\boldsymbol{x}) - \langle \boldsymbol{d}^*, \boldsymbol{f}(\boldsymbol{x}) \rangle \\ \text{s.t.} \quad & \boldsymbol{D}^* \geq \boldsymbol{d}^* \geq \boldsymbol{0} \end{aligned} \tag{C.3}$$

The additional constraint $\boldsymbol{D}^* \geq \boldsymbol{d}^*$ is redundant as it is non-binding in the optimum. Then, interchanging the 'inf' and 'sup' operators we obtain

$$\begin{aligned} \sup_{\boldsymbol{x}} \inf_{\boldsymbol{d}_t^*} \; & \rho(\boldsymbol{x}) - \langle \boldsymbol{d}^*, \boldsymbol{f}(\boldsymbol{x}) \rangle &= \sup_{\boldsymbol{x}} \rho(\boldsymbol{x}) - \langle \boldsymbol{D}^*, [\boldsymbol{f}(\boldsymbol{x})]^+ \rangle. \quad \text{(C.4)} \\ \text{s.t.} \; & \boldsymbol{D}^* \geq \boldsymbol{d}^* \geq \boldsymbol{0}, \end{aligned}$$

Note that (C.4) is equivalent to (C.3) by strong duality.[1] In summary, each of the optimization problems (C.2) through (C.4) has the same optimal value as the original problem (B.1). This observation completes the proof. □

Let us next discuss maximization problems of the form (B.7) with both inequality and equality constraints. Arguing as in the proof of proposition C.1, one can show that (B.7) is equivalent to

$$\begin{aligned} \sup_{\boldsymbol{x}} \; \rho(\boldsymbol{x}) - \langle \boldsymbol{D}^{\text{in}*}, [\boldsymbol{f}^{\text{in}}(\boldsymbol{x})]^+ \rangle - \langle \boldsymbol{D}^{\text{eq}*+}, [\boldsymbol{f}^{\text{eq}}(\boldsymbol{x})]^+ \rangle \quad \text{(C.5)} \\ - \langle \boldsymbol{D}^{\text{eq}*-}, [\boldsymbol{f}^{\text{eq}}(\boldsymbol{x})]^- \rangle. \end{aligned}$$

Thereby, $\boldsymbol{D}^{\text{in}*} \in \mathbb{R}^{\text{in}}$ and $\boldsymbol{D}^{\text{eq}*+}, \boldsymbol{D}^{\text{eq}*-} \in \mathbb{R}^{\text{eq}}$ are arbitrary bounding vectors which satisfy the inequalities

$$\boldsymbol{D}^{\text{in}*} \geq \boldsymbol{d}^{\text{in}*}, \quad \boldsymbol{D}^{\text{eq}*+} \geq [\boldsymbol{d}^{\text{eq}*}]^+, \quad \text{and} \quad \boldsymbol{D}^{\text{eq}*-} \geq [\boldsymbol{d}^{\text{eq}*}]^-$$

for some solution $(\boldsymbol{d}^{\text{in}*}, \boldsymbol{d}^{\text{eq}*})$ of the dual problem (B.11). To simplify notation, one can reformulate (B.7) as

$$\sup_{\boldsymbol{x}} \{\rho(\boldsymbol{x}) \,|\, \boldsymbol{f}(\boldsymbol{x}) \leq \boldsymbol{0}\} \quad \text{with} \quad \boldsymbol{f} := (\boldsymbol{f}^{\text{in}}, \boldsymbol{f}^{\text{eq}}, -\boldsymbol{f}^{\text{eq}}) \qquad \text{(C.6)}$$

and (C.5) as

$$\sup_{\boldsymbol{x}} \; \rho(\boldsymbol{x}) - \langle \boldsymbol{D}^*, [\boldsymbol{f}(\boldsymbol{x})]^+ \rangle \quad \text{with} \quad \boldsymbol{D}^* := (\boldsymbol{D}^{\text{in}*}, \boldsymbol{D}^{\text{eq}*+}, \boldsymbol{D}^{\text{eq}*-}). \quad \text{(C.7)}$$

Formally speaking, we may state the following corollary C.2, which generalizes proposition C.1 by allowing for equality constraints.

Corollary C.2. *The mathematical programs (C.6) and (C.7) have the same optimal values.*

The general findings of this appendix are related to results on *exact penalty functions* for nonlinear programs [115]. A version of proposition C.1 for linear programs can be found in [74].

---

[1] Consider the embedding

$$P(\boldsymbol{x}, \boldsymbol{d}) := \rho(\boldsymbol{x}) - \langle \boldsymbol{D}^*, [\boldsymbol{f}(\boldsymbol{x}) - \boldsymbol{d}]^+ \rangle$$

and use an elementary argument to show that the corresponding value function $\Phi(\boldsymbol{d}) := \sup_{\boldsymbol{x}} P(\boldsymbol{x}, \boldsymbol{d})$ is subdifferentiable on the entire space. Thus, strong duality follows.

D

Parametric Families of Linear Functions

In this appendix we derive a useful result about convex mappings which are linear affine in some of their arguments. Proposition D.1 below helps us to gain deeper insights into the properties of an important class of constraint functions considered in section 5.2.

Proposition D.1. *Consider a vector-valued mapping* $\boldsymbol{w} : \mathbb{R}^L \to \mathbb{R}^n$ *and a real-valued mapping* $\kappa : \mathbb{R}^L \to \mathbb{R}$*. Moreover, assume that the function*

$$f : \begin{cases} \mathbb{R}^n \times \mathbb{R}^L \to \mathbb{R} \\ (\boldsymbol{x}, \boldsymbol{\xi}) \mapsto \langle \boldsymbol{w}(\boldsymbol{\xi}), \boldsymbol{x} \rangle + \kappa(\boldsymbol{\xi}) \end{cases}$$

is finite and jointly convex in $(\boldsymbol{x}, \boldsymbol{\xi})$ *on a convex neighborhood of* $(\bar{\boldsymbol{x}}, \bar{\boldsymbol{\xi}})$*. Then,* $\boldsymbol{w}$ *is constant and* κ *is convex on a convex neighborhood of* $\bar{\boldsymbol{\xi}}$*.*

Proof. Without loss of generality we may assume that $(\bar{\boldsymbol{x}}, \bar{\boldsymbol{\xi}}) = (\mathbf{0}, \mathbf{0})$.[1] Then, there are a tolerance $\varepsilon > 0$ and two compact cubes

$$U := \{\boldsymbol{x} \in \mathbb{R}^n \,|\, \|\boldsymbol{x}\|_\infty \leq \varepsilon\} \quad \text{and} \quad V := \{\boldsymbol{\xi} \in \mathbb{R}^L \,|\, \|\boldsymbol{\xi}\|_\infty \leq \varepsilon\}$$

such that f is finite and jointly convex in $(\boldsymbol{x}, \boldsymbol{\xi})$ on a convex neighborhood of $U \times V$. Here, $\|\cdot\|_\infty$ stands for the vector ∞-norm. By fixing $\boldsymbol{x} \equiv \mathbf{0}$, it can easily be seen that convexity of f requires κ to be convex on a convex neighborhood of V. Moreover, $\boldsymbol{w}$ and κ are necessarily Lipschitz continuous on V. For the

[1] If $(\bar{\boldsymbol{x}}, \bar{\boldsymbol{\xi}}) \neq (\mathbf{0}, \mathbf{0})$, we may set $\boldsymbol{x}' := \boldsymbol{x} - \bar{\boldsymbol{x}}$ and $\boldsymbol{\xi}' := \boldsymbol{\xi} - \bar{\boldsymbol{\xi}}$. Moreover, define

$$f'(\boldsymbol{x}', \boldsymbol{\xi}') := \langle \boldsymbol{w}'(\boldsymbol{\xi}'), \boldsymbol{x}' \rangle + \kappa'(\boldsymbol{\xi}'),$$

where $\boldsymbol{w}'(\boldsymbol{\xi}') := \boldsymbol{w}(\boldsymbol{\xi}' + \bar{\boldsymbol{\xi}})$ and $\kappa'(\boldsymbol{\xi}') := \kappa(\boldsymbol{\xi}' + \bar{\boldsymbol{\xi}}) + \langle \boldsymbol{w}(\boldsymbol{\xi}' + \bar{\boldsymbol{\xi}}), \bar{\boldsymbol{x}} \rangle$. This f' is finite and convex on a neighborhood of the origin. If we can show that $\boldsymbol{w}'$ is constant and κ' is convex on a neighborhood of $\boldsymbol{\xi}' = \mathbf{0}$, then we find immediately that $\boldsymbol{w}$ is constant and κ is convex on a neighborhood of $\boldsymbol{\xi} = \bar{\boldsymbol{\xi}}$.

further argumentation it is convenient to introduce an extended-real-valued function which is convex on the entire underlying space.

$$\hat{f}(\boldsymbol{x},\boldsymbol{\xi}) = \begin{cases} f(\boldsymbol{x},\boldsymbol{\xi}) & \text{for } (\boldsymbol{x},\boldsymbol{\xi}) \in U \times V, \\ +\infty & \text{else.} \end{cases}$$

By construction, $\hat{f}$ is a proper convex function on $\mathbb{R}^n \times \mathbb{R}^L$. Let us then calculate the convex conjugate of $\hat{f}$ with respect to the first argument. We find

$$\begin{aligned} \hat{f}^*(\boldsymbol{x}^*,\boldsymbol{\xi}) &= \sup_{\boldsymbol{x}} \langle \boldsymbol{x}^*, \boldsymbol{x}\rangle - \hat{f}(\boldsymbol{x},\boldsymbol{\xi}) \\ &= \begin{cases} \varepsilon \, \|\boldsymbol{x}^* - \boldsymbol{w}(\boldsymbol{\xi})\|_1 - \kappa(\boldsymbol{\xi}) & \text{for } \boldsymbol{\xi} \in V, \\ -\infty & \text{else,} \end{cases} \end{aligned}$$

where $\|\cdot\|_1$ stands for the vector 1-norm. Since $\hat{f}$ is convex on $\mathbb{R}^n \times \mathbb{R}^L$, the conjugate $\hat{f}^*$ is convex in $\boldsymbol{x}^*$ for all $\boldsymbol{\xi} \in \mathbb{R}^L$ and concave in $\boldsymbol{\xi}$ for all $\boldsymbol{x}^* \in \mathbb{R}^n$. Assume now that $\boldsymbol{w}$ is not constant on V. Then, there are two vectors $\boldsymbol{\xi}_0, \boldsymbol{\xi}_1 \in V$ such that $\boldsymbol{w}(\boldsymbol{\xi}_0) \neq \boldsymbol{w}(\boldsymbol{\xi}_1)$. Moreover, we may introduce a linear affine function

$$\tilde{\boldsymbol{\xi}} : [0,1] \to V \quad \text{such that} \quad \tilde{\boldsymbol{\xi}}(0) = \boldsymbol{\xi}_0 \quad \text{and} \quad \tilde{\boldsymbol{\xi}}(1) = \boldsymbol{\xi}_1.$$

By construction, $\boldsymbol{w} \circ \tilde{\boldsymbol{\xi}}$ and $\kappa \circ \tilde{\boldsymbol{\xi}}$ are Lipschitz continuous, and the mapping

$$\varphi(s) := \hat{f}^*(\boldsymbol{x}^*, \tilde{\boldsymbol{\xi}}(s)) = \varepsilon \, \|\boldsymbol{x}^* - \boldsymbol{w} \circ \tilde{\boldsymbol{\xi}}(s)\|_1 - \kappa \circ \tilde{\boldsymbol{\xi}}(s)$$

is concave on $[0,1]$ for all $\boldsymbol{x}^* \in \mathbb{R}^n$. Denote by D the set of all $s \in [0,1]$ where both $\boldsymbol{w} \circ \tilde{\boldsymbol{\xi}}$ and $\kappa \circ \tilde{\boldsymbol{\xi}}$ are differentiable. The Lipschitz continuity guarantees via the theorem of Rademacher (see e.g. [36] for a modern proof) that D has Lebesgue measure 1. In addition, we may write

$$\boldsymbol{w} \circ \tilde{\boldsymbol{\xi}}(s) = \int_0^s \dot{\boldsymbol{w}}(u) du + \boldsymbol{w}(\boldsymbol{\xi}_0),$$

where $\dot{\boldsymbol{w}}(s)$ is the first derivative of $\boldsymbol{w} \circ \tilde{\boldsymbol{\xi}}$ at s whenever $s \in D$ and zero otherwise. Analogously, we find

$$\kappa \circ \tilde{\boldsymbol{\xi}}(s) = \int_0^s \dot{\kappa}(u) du + \kappa(\boldsymbol{\xi}_0),$$

where $\dot{\kappa}(s)$ is the first derivative of $\kappa \circ \tilde{\boldsymbol{\xi}}$ at s for $s \in D$ and zero otherwise. Next, define D_0 as the set of all $s \in D$ where $\dot{\boldsymbol{w}}(s) \neq 0$. As $\boldsymbol{w}(\boldsymbol{\xi}_0) \neq \boldsymbol{w}(\boldsymbol{\xi}_1)$, we may conclude that D_0 has strictly positive Lebesgue measure. A fortiori, D_0 is non-empty. Choose $s_0 \in D_0$, and set $\boldsymbol{x}^* = \boldsymbol{w} \circ \tilde{\boldsymbol{\xi}}(s_0)$. Then, the mapping φ can be expanded around s_0.

$$\varphi(s) - \varphi(s_0) = \varepsilon \, \|\dot{\boldsymbol{w}}(s_0)\|_1 \, |s - s_0| - \dot{\kappa}(s_0)\,(s - s_0) + o(s - s_0)$$

By construction, $\|\dot{\boldsymbol{w}}(s_0)\|_1$ is strictly positive, which contradicts concavity of φ. Thus, our assumption is false, and $\boldsymbol{w}$ is constant on V. □

The assertions of proposition D.1 can be proved much more directly if it is assumed that $\boldsymbol{w}$ and κ are twice continuously differentiable on a convex neighborhood of $\bar{\boldsymbol{\xi}}$. In fact, by calculating the Hessian of f at an arbitrary point $(\boldsymbol{x}, \boldsymbol{\xi}) \in U \times V$, one can easily verify that the first derivative of $\boldsymbol{w}$ vanishes and the Hessian of κ is positive semidefinite. Otherwise the Hessian of f would not be positive semidefinite at $(\boldsymbol{x}, \boldsymbol{\xi})$ implying that f would not be (locally) convex.

E

Lipschitz Continuity of sup-Projections

In this appendix we study the parametric dependence of the Lagrange multipliers associated with a family of maximization problems. To this end, we need a specific result about Lipschitz continuity of sup-projections. This technical result is needed for the proof of proposition 5.12 in the main text. For the sake of transparent notation, here, we consider a simple parametric optimization problem over the Euclidean space $\mathbb{R}^n$

$$\begin{aligned} \Phi(\boldsymbol{\eta}, \boldsymbol{\xi}) = \sup_{\boldsymbol{x} \in \mathbb{R}^n} \; & \rho(\boldsymbol{x}, \boldsymbol{\eta}, \boldsymbol{\xi}) \\ \text{s.t.} \quad & \boldsymbol{f}(\boldsymbol{x}, \boldsymbol{\xi}) \leq \boldsymbol{0} \end{aligned} \tag{E.1}$$

which depends on the parameters $\boldsymbol{\eta} \in \mathbb{R}^K$ and $\boldsymbol{\xi} \in \mathbb{R}^L$. The extended-real-valued objective function ρ is assumed to be concave in $\boldsymbol{x}$. Moreover, the vector-valued constraint functions

$$\begin{aligned} \boldsymbol{f}^{\text{in}} &: \mathbb{R}^n \times \mathbb{R}^L \to \mathbb{R}^{r^{\text{in}}} \\ \boldsymbol{f}^{\text{eq}} &: \mathbb{R}^n \times \mathbb{R}^L \to \mathbb{R}^{r^{\text{eq}}} \end{aligned}$$

corresponding to inequality and equality constraints are continuous, and the combination

$$\boldsymbol{f} := (\boldsymbol{f}^{\text{in}}, \boldsymbol{f}^{\text{eq}}, -\boldsymbol{f}^{\text{eq}}) : \mathbb{R}^n \times \mathbb{R}^L \to \mathbb{R}^r$$

is convex in $\boldsymbol{x}$, where $r = r^{\text{in}} + 2 \cdot r^{\text{eq}}$. The constraint functions define a closed-valued feasible set mapping

$$\boldsymbol{\xi} \mapsto X(\boldsymbol{\xi}) := \{\boldsymbol{x} \,|\, \boldsymbol{f}(\boldsymbol{x}, \boldsymbol{\xi}) \leq \boldsymbol{0}\}.$$

Next, let $\Theta \subset \mathbb{R}^K$ and $\Xi \subset \mathbb{R}^L$ be compact (not necessarily convex) sets, and set $Z := \Theta \times \Xi$. Denote by Y the graph of the feasible set mapping X over Z. Then, we require ρ to be a subdifferentiable saddle function on a convex neighborhood of Y being convex in $\boldsymbol{\eta}$ and jointly concave in $(\boldsymbol{x}, \boldsymbol{\xi})$. Moreover,

$\boldsymbol{f}$ is regularizable in the following sense: there is a continuous vector-valued function $\boldsymbol{\kappa} : \mathbb{R}^L \to \mathbb{R}^r$ which is constant in $\boldsymbol{x}$ and $\boldsymbol{\eta}$, and both $\boldsymbol{\kappa}$ and $\boldsymbol{f} + \boldsymbol{\kappa}$ are convex functions on a convex neighborhood of Y. It is convenient to write the mapping $\boldsymbol{\kappa}$ as

$$\boldsymbol{\kappa} = (\boldsymbol{\kappa}^{\text{in}}, \boldsymbol{\kappa}^{\text{eq}+}, \boldsymbol{\kappa}^{\text{eq}-}) \quad \text{where} \quad \begin{cases} \boldsymbol{\kappa}^{\text{in}} \;\; : \mathbb{R}^L \to \mathbb{R}^{r^{\text{in}}}, \\ \boldsymbol{\kappa}^{\text{eq}+} : \mathbb{R}^L \to \mathbb{R}^{r^{\text{eq}}}, \\ \boldsymbol{\kappa}^{\text{eq}-} : \mathbb{R}^L \to \mathbb{R}^{r^{\text{eq}}}. \end{cases}$$

This representation reflects the grouping of $\boldsymbol{f} = (\boldsymbol{f}^{\text{in}}, \boldsymbol{f}^{\text{eq}}, -\boldsymbol{f}^{\text{eq}})$ by inequality and equality constraints. By applying proposition D.1 in appendix D to each component of the regularized constraint function $\boldsymbol{f}^{\text{eq}} + \boldsymbol{\kappa}^{\text{eq}+}$ and for every reference point $(\boldsymbol{x}, \boldsymbol{\eta}, \boldsymbol{\xi}) \in Y$, we may conclude that

$$\boldsymbol{f}^{\text{eq}}(\boldsymbol{x}, \boldsymbol{\xi}) \equiv W\boldsymbol{x} - \boldsymbol{h}(\boldsymbol{\xi}),$$

on a neighborhood of Y. Thereby, the matrix W is constant, and the mapping $\boldsymbol{h}$ can be represented as a difference of two convex functions. Without loss of generality one may assume that $\boldsymbol{h} = \boldsymbol{\kappa}^{\text{eq}+} - \boldsymbol{\kappa}^{\text{eq}-}$. Finally, we postulate that the parametric maximization problem (E.1) satisfies Slater's constraint qualification for any $(\boldsymbol{\eta}, \boldsymbol{\xi}) \in Z$, and the multifunction X is bounded on a neighborhood of Z.

In order to study the Lagrange multipliers associated with the explicit constraints in (E.1), we should first consider a family of perturbed maximization problems

$$\begin{aligned} \Phi'(\boldsymbol{d}^{\text{in}}, \boldsymbol{d}^{\text{eq}}, \boldsymbol{\eta}, \boldsymbol{\xi}) = \; & \sup_{\boldsymbol{x} \in \mathbb{R}^n} \; \rho(\boldsymbol{x}, \boldsymbol{\eta}, \boldsymbol{\xi}) \\ & \text{s.t.} \quad \boldsymbol{f}^{\text{in}}(\boldsymbol{x}, \boldsymbol{\xi}) \le \boldsymbol{d}^{\text{in}} \\ & \qquad\;\; \boldsymbol{f}^{\text{eq}}(\boldsymbol{x}, \boldsymbol{\xi}) = \boldsymbol{d}^{\text{eq}} \end{aligned} \tag{E.2}$$

which depends on the perturbation parameters $\boldsymbol{d}^{\text{in}} \in \mathbb{R}^{r^{\text{in}}}$ and $\boldsymbol{d}^{\text{eq}} \in \mathbb{R}^{r^{\text{eq}}}$ (see appendix B for an introduction to Lagrangian duality). It is useful to introduce a new feasible set mapping

$$(\boldsymbol{d}^{\text{in}}, \boldsymbol{d}^{\text{eq}}, \boldsymbol{\xi}) \mapsto X'(\boldsymbol{d}^{\text{in}}, \boldsymbol{d}^{\text{eq}}, \boldsymbol{\xi}) := \{\boldsymbol{x} \,|\, \boldsymbol{f}^{\text{in}}(\boldsymbol{x}, \boldsymbol{\xi}) \le \boldsymbol{d}^{\text{in}}, \boldsymbol{f}^{\text{eq}}(\boldsymbol{x}, \boldsymbol{\xi}) = \boldsymbol{d}^{\text{eq}}\}$$

which depends both on $\boldsymbol{\xi}$ and the perturbation parameters. Next, define

$$Z' := \{(\boldsymbol{0}, \boldsymbol{0})\} \times \Theta \times \Xi \subset \mathbb{R}^{r^{\text{in}}} \times \mathbb{R}^{r^{\text{eq}}} \times \mathbb{R}^K \times \mathbb{R}^L.$$

Moreover, denote by Y' the graph of the feasible set mapping X' over Z'. For the further argumentation we need some preliminary results about the new feasible set mapping.

Proposition E.1. *The feasible set mapping X' is non-empty-valued, bounded, and usc on a neighborhood of Z'.*

Proof. Slater's constraint qualification implies via proposition 3.2 that the feasible set mapping X' is non-empty-valued on a neighborhood of Z'. Next, we prove boundedness. Let us introduce an auxiliary constraint function

$$g(\boldsymbol{d}^{\text{in}}, \boldsymbol{d}^{\text{eq}}, \boldsymbol{x}, \boldsymbol{\xi}) := \max\{\max\{f_i^{\text{in}}(\boldsymbol{x}, \boldsymbol{\xi}) - d_i^{\text{in}}; i = 1, \ldots, r^{\text{in}}\},$$
$$\max\{|f_i^{\text{eq}}(\boldsymbol{x}, \boldsymbol{\xi}) - d_i^{\text{eq}}|; i = 1, \ldots, r^{\text{eq}}\}\},$$

which is continuous and convex in $\boldsymbol{x}$ on the entire underlying space. Due to boundedness of X on a neighborhood of Z, there is a compact neighborhood V' of Ξ and a positive real number R such that

$$X(\boldsymbol{\xi}) = \{\boldsymbol{x} \,|\, g(\boldsymbol{0}, \boldsymbol{0}, \boldsymbol{x}, \boldsymbol{\xi}) \leq 0\} \subset B_R \quad \forall \boldsymbol{\xi} \in V', \tag{E.3}$$

where $B_R := \{\boldsymbol{x} \,|\, \|\boldsymbol{x}\| < R\}$ represents the open ball of radius R around the origin. Then, the inclusion (E.3) and continuity of g imply that

$$g(\boldsymbol{0}, \boldsymbol{0}, \boldsymbol{x}, \boldsymbol{\xi}) \geq \varepsilon > 0 \quad \forall (\boldsymbol{x}, \boldsymbol{\xi}) \in S_R \times V',$$

where $S_R := \{\boldsymbol{x} \,|\, \|\boldsymbol{x}\| = R\}$, and ε is some strictly positive real number. By construction, the auxiliary constraint function g is uniformly continuous on any bounded set. Thus, there exists a neighborhood U' of $(\boldsymbol{d}^{\text{in}}, \boldsymbol{d}^{\text{eq}}) = (\boldsymbol{0}, \boldsymbol{0})$ such that

$$g(\boldsymbol{d}^{\text{in}}, \boldsymbol{d}^{\text{eq}}, \boldsymbol{x}, \boldsymbol{\xi}) > 0 \quad \forall (\boldsymbol{d}^{\text{in}}, \boldsymbol{d}^{\text{eq}}) \in U', (\boldsymbol{x}, \boldsymbol{\xi}) \in S_R \times V'.$$

As implied by proposition 3.2, the feasible set mapping X' has locally a continuous selector on Z'. Consequently, there exists a neighborhood V'' of Ξ and a neighborhood U'' of $(\boldsymbol{d}^{\text{in}}, \boldsymbol{d}^{\text{eq}}) = (\boldsymbol{0}, \boldsymbol{0})$ such that

$$X'(\boldsymbol{d}^{\text{in}}, \boldsymbol{d}^{\text{eq}}, \boldsymbol{\xi}) \cap B_R \neq \emptyset \quad \forall (\boldsymbol{d}^{\text{in}}, \boldsymbol{d}^{\text{eq}}) \in U'', \boldsymbol{\xi} \in V''.$$

Setting $U := U' \cap U''$, $V := V' \cap V''$, and by using convexity of g in the decision variables we may conclude that

$$X'(\boldsymbol{d}^{\text{in}}, \boldsymbol{d}^{\text{eq}}, \boldsymbol{\xi}) = \{\boldsymbol{x} \,|\, g(\boldsymbol{d}^{\text{in}}, \boldsymbol{d}^{\text{eq}}, \boldsymbol{x}, \boldsymbol{\xi}) \leq 0\} \subset B_R \quad \forall (\boldsymbol{d}^{\text{in}}, \boldsymbol{d}^{\text{eq}}) \in U, \boldsymbol{\xi} \in V.$$

This observation proves boundedness of the feasible set mapping X' on a neighborhood of Z'. Moreover, X' has a closed graph since the constraint functions are continuous on the entire space. Upper semicontinuity of X' on a neighborhood of Z' then follows from [13, proposition 11.9(b)]. □

Corollary E.2. *For any neighborhood U of Y' there is a neighborhood V of Z' such that the graph of X' over V is a subset of U.*

Proof. Use upper semicontinuity of X' as in proposition 3.10. □

Proposition E.3. *There are open bounded (not necessarily convex) sets*

$$\left.\begin{array}{lll} Y_\cap & \subset & \mathbb{R}^{r^{\text{in}}} \times \mathbb{R}^{r^{\text{eq}}} \times \mathbb{R}^n \times \mathbb{R}^L \\ Y_\cup & \subset & \mathbb{R}^K \\ Z_\cap & \subset & \mathbb{R}^{r^{\text{in}}} \times \mathbb{R}^{r^{\text{eq}}} \times \mathbb{R}^L \\ Z_\cup & \subset & \mathbb{R}^K \end{array}\right\} \tag{E.4}$$

with the following properties:

(a) $Y_\cap \times Y_\cup$ *is a neighborhood of* Y'*;*

(b) $Z_\cap \times Z_\cup$ *is a neighborhood of* Z'*;*

(c) ρ *is a continuous saddle function on* $\operatorname{co} Y_\cap \times \operatorname{co} Y_\cup$ *being concave in* $\boldsymbol{x}$*, convex in* $\boldsymbol{\eta}$*, and constant in* $(\boldsymbol{d}^{\text{in}}, \boldsymbol{d}^{\text{eq}}, \boldsymbol{\xi})$*;*

(d) both $\boldsymbol{f} + \boldsymbol{\kappa}$ *and* $\boldsymbol{\kappa}$ *are continuous convex functions on* $\operatorname{co} Y_\cap \times \operatorname{co} Y_\cup$*;*

(e) the graph of X' *over* $Z_\cap \times Z_\cup$ *is a subset of* $Y_\cap \times Y_\cup$*;*

(f) the multifunction X' *is non-empty compact-valued on* $Z_\cap \times Z_\cup$*.*

Proof. By the general assumptions, it is easy to find open bounded sets $Y_\cap$ and $Y_\cup$ satisfying the requirements (a), (c), and (d). Furthermore, by proposition E.1 and corollary E.2 it is possible to find open bounded sets $Z_\cap$ and $Z_\cup$ which fulfill the remaining requirements (b), (e), and (f). □

Proposition E.4. *The value function* Φ' *is Lipschitz continuous on a neighborhood of* Z'*.*

Proof. Define an auxiliary objective function

$$q(\boldsymbol{x}, \boldsymbol{\eta}, \boldsymbol{\xi}) := \begin{cases} \rho(\boldsymbol{x}, \boldsymbol{\eta}, \boldsymbol{\xi}) & \text{on} \quad \operatorname{co} Y_\cap \times \operatorname{co} Y_\cup, \\ +\infty & \text{on} \quad \operatorname{co} Y_\cap \times (\operatorname{co} Y_\cup)^{\text{c}}, \\ -\infty & \text{everywhere else.} \end{cases}$$

By assertion (c) of proposition E.3, q is a saddle function on the entire underlying space. Moreover, introduce two auxiliary value functions.

$$\begin{aligned} Q(\boldsymbol{d}^{\text{in}}, \boldsymbol{d}^{\text{eq}}, \boldsymbol{\eta}, \boldsymbol{\xi}) = \sup_{\boldsymbol{x} \in \mathbb{R}^n} \; & q(\boldsymbol{x}, \boldsymbol{\eta}, \boldsymbol{\xi}) \\ \text{s.t.} \quad & \boldsymbol{f}^{\text{in}}(\boldsymbol{x}, \boldsymbol{\xi}) \leq \boldsymbol{d}^{\text{in}} \\ & \boldsymbol{f}^{\text{eq}}(\boldsymbol{x}, \boldsymbol{\xi}) = \boldsymbol{d}^{\text{eq}} \end{aligned} \tag{E.5a}$$

$$\begin{aligned} Q'(\boldsymbol{d}^{\text{in}}, \boldsymbol{d}^{\text{eq}}, \boldsymbol{\eta}, \boldsymbol{\xi}) = \sup_{\boldsymbol{x} \in \mathbb{R}^n} \; & q(\boldsymbol{x}, \boldsymbol{\eta}, \boldsymbol{\xi}) \\ \text{s.t.} \quad & \boldsymbol{f}^{\text{in},\prime}(\boldsymbol{x}, \boldsymbol{\xi}) \leq \boldsymbol{d}^{\text{in}} \\ & \boldsymbol{f}^{\text{eq},\prime}(\boldsymbol{x}, \boldsymbol{\xi}) = \boldsymbol{d}^{\text{eq}} \end{aligned} \tag{E.5b}$$

Notice that Q' closely resembles Q. However, the potentially nonconvex constraint functions in (E.5a) have been replaced in (E.5b) by

$$\boldsymbol{f}^{\text{in},\prime} := \boldsymbol{f}^{\text{in}} + \boldsymbol{\kappa}^{\text{in}} \quad \text{and} \quad \boldsymbol{f}^{\text{eq},\prime} := \boldsymbol{f}^{\text{eq}} + \boldsymbol{\kappa}^{\text{eq}+} - \boldsymbol{\kappa}^{\text{eq}-}.$$

By construction, $\boldsymbol{f}^{\text{in},\prime}$ is jointly convex in $\boldsymbol{x}$ and $\boldsymbol{\xi}$ whereas $\boldsymbol{f}^{\text{eq},\prime}$ is linear affine in $\boldsymbol{x}$ and constant in $\boldsymbol{\xi}$ on $Y_\cap \times Y_\cup$. Next, define a homeomorphism

$$\iota : (\boldsymbol{d}^{\text{in}}, \boldsymbol{d}^{\text{eq}}, \boldsymbol{\eta}, \boldsymbol{\xi}) \mapsto (\boldsymbol{d}^{\text{in}} + \boldsymbol{\kappa}^{\text{in}}(\boldsymbol{\xi}), \boldsymbol{d}^{\text{eq}} + \boldsymbol{\kappa}^{\text{eq}+}(\boldsymbol{\xi}) - \boldsymbol{\kappa}^{\text{eq}-}(\boldsymbol{\xi}), \boldsymbol{\eta}, \boldsymbol{\xi}).$$

It can easily be verified that ι relates the auxiliary value functions (E.5) through $Q = Q' \circ \iota$. This representation will be used to prove that Q is locally Lipschitz continuous[1] on $Z_\cap \times Z_\cup$. By the properties of the objective and constraint functions in (E.5b), the perturbed value function Q' is convex in $\boldsymbol{\eta}$ and jointly concave in $(\boldsymbol{d}^{\text{in}}, \boldsymbol{d}^{\text{eq}}, \boldsymbol{\xi})$ on the entire underlying space (cf. also the related argument in the proof of theorem 3.14). Moreover, by the assertions (e) and (f) of proposition E.3, the parametric optimization problem (E.5a) is feasible for any parameter $(\boldsymbol{d}^{\text{in}}, \boldsymbol{d}^{\text{eq}}, \boldsymbol{\eta}, \boldsymbol{\xi}) \in Z_\cap \times Z_\cup$. Then, the Weierstrass maximum principle proves that Q is pointwise finite on $Z_\cap \times Z_\cup$. This in turn implies pointwise finiteness of Q' on $\iota(Z_\cap \times Z_\cup)$, which is open, but generally nonconvex. As Q' is a saddle function, we may invoke [89, theorem 35.1] to show that Q' is locally Lipschitz on $\iota(Z_\cap \times Z_\cup)$. By convexity of the mapping $\boldsymbol{\kappa}$, the homeomorphism ι is locally Lipschitz on $Z_\cap \times Z_\cup$. This is a consequence of [89, theorem 24.7]. Hence, the composition $Q = Q' \circ \iota$ is locally Lipschitz on the open set $Z_\cap \times Z_\cup$.

By assertion (e) of proposition E.3, the optimal value function Φ' coincides with Q on $Z_\cap \times Z_\cup$. Thus, Φ' is locally Lipschitz on $Z_\cap \times Z_\cup$ and globally Lipschitz on every compact subset of $Z_\cap \times Z_\cup$. This observation completes the proof. □

Proposition E.5. *The Lagrange multipliers associated with the explicit constraints in the parametric maximization problem (E.1) are uniformly bounded for all* $(\boldsymbol{\eta}, \boldsymbol{\xi})$ *in a neighborhood of* Z.

Proof. As argued in proposition B.3, for any fixed parameter $(\boldsymbol{\eta}, \boldsymbol{\xi}) \in Z$ the dual feasible set of problem (E.1) is convex and unbounded. In contrast, the dual solution set, i.e. the set of Lagrange multipliers, is compact and reduces to the subdifferential of the perturbed value function $\Phi'(\cdot, \cdot, \boldsymbol{\eta}, \boldsymbol{\xi})$ evaluated at $(\boldsymbol{d}^{\text{in}}, \boldsymbol{d}^{\text{eq}}) = (\mathbf{0}, \mathbf{0})$.

$$D^*_{\text{opt}}(\boldsymbol{\eta}, \boldsymbol{\xi}) := \partial_{\boldsymbol{d}^{\text{in}}} \times \partial_{\boldsymbol{d}^{\text{eq}}} \Phi'(\boldsymbol{d}^{\text{in}}, \boldsymbol{d}^{\text{eq}}, \boldsymbol{\eta}, \boldsymbol{\xi})\big|_{(\boldsymbol{d}^{\text{in}}, \boldsymbol{d}^{\text{eq}}) = (\mathbf{0}, \mathbf{0})}$$

---

[1] The Lipschitzian properties of general inf-projections are analyzed by Wets [109]. Here, however, Lipschitz continuity of the sup-projection Q can directly be established by exploiting the regularizability of the constraint functions and the saddle structure of q.

By proposition E.4, the value function Φ' is Lipschitz continuous on a neighborhood of Z' with some constant Λ. Thus, the following estimates hold

$$\left.\begin{array}{l} \boldsymbol{D}^{\mathrm{in}*} \geq \boldsymbol{d}^{\mathrm{in}*} \geq \quad \boldsymbol{0} \\ \boldsymbol{D}^{\mathrm{eq}*+} \geq \boldsymbol{d}^{\mathrm{eq}*} \geq -\boldsymbol{D}^{\mathrm{eq}*-} \end{array}\right\} \quad \forall\,(\boldsymbol{d}^{\mathrm{in}*}, \boldsymbol{d}^{\mathrm{eq}*}) \in D^*_{\mathrm{opt}}(\boldsymbol{\eta}, \boldsymbol{\xi}), \tag{E.6}$$

where

$$\boldsymbol{D}^{\mathrm{in}*} := \underbrace{(\Lambda, \ldots, \Lambda)}_{r^{\mathrm{in}} \text{ times}} \quad \text{and} \quad \boldsymbol{D}^{\mathrm{eq}*+} := \boldsymbol{D}^{\mathrm{eq}*-} := \underbrace{(\Lambda, \ldots, \Lambda)}_{r^{\mathrm{eq}} \text{ times}}.$$

Consequently, the set of optimal dual solutions $D^*_{\mathrm{opt}}(\boldsymbol{\eta}, \boldsymbol{\xi})$ is uniformly bounded on a neighborhood of Z. Thus, the claim follows. □

Finally, we state a technical result which is needed in the proof of proposition 5.12 in the main text. Assume that $\bar{\rho}$ is an alternative objective function valued in the extended reals. Moreover, assume that $\bar{\rho}$ is concave in the decision vector $\boldsymbol{x}$, and consider the parametric optimization problem

$$\begin{aligned} \bar{\Phi}(\boldsymbol{\eta}, \boldsymbol{\xi}) = \sup_{\boldsymbol{x} \in \mathbb{R}^n} & \ \bar{\rho}(\boldsymbol{x}, \boldsymbol{\eta}, \boldsymbol{\xi}) \\ \text{s.t.} & \ \boldsymbol{f}(\boldsymbol{x}, \boldsymbol{\xi}) \leq \boldsymbol{0}. \end{aligned} \tag{E.7}$$

Proposition E.6. *Assume that $\rho = \bar{\rho}$ on a neighborhood of Y. Then, the Lagrange multiplier sets of the problems (E.1) and (E.7) coincide at any reference point $(\boldsymbol{\eta}, \boldsymbol{\xi}) \in Z$.*

Proof. In order to study the Lagrange multipliers associated with the explicit constraints in (E.7), we have to consider the usual family of perturbed maximization problems

$$\begin{aligned} \bar{\Phi}'(\boldsymbol{d}^{\mathrm{in}}, \boldsymbol{d}^{\mathrm{eq}}, \boldsymbol{\eta}, \boldsymbol{\xi}) = \sup_{\boldsymbol{x} \in \mathbb{R}^n} & \ \bar{\rho}(\boldsymbol{x}, \boldsymbol{\eta}, \boldsymbol{\xi}) \\ \text{s.t.} & \ \boldsymbol{f}^{\mathrm{in}}(\boldsymbol{x}, \boldsymbol{\xi}) \leq \boldsymbol{d}^{\mathrm{in}} \\ & \ \boldsymbol{f}^{\mathrm{eq}}(\boldsymbol{x}, \boldsymbol{\xi}) = \boldsymbol{d}^{\mathrm{eq}}. \end{aligned}$$

By corollary E.2 we may conclude that $\bar{\Phi}' = \Phi'$ on a neighborhood of Z'. Notice that both $\bar{\Phi}'$ and Φ' are concave in $(\boldsymbol{x}, \boldsymbol{d}^{\mathrm{in}}, \boldsymbol{d}^{\mathrm{eq}})$ on the entire space. Thus, we find

$$\partial_{\boldsymbol{d}^{\mathrm{in}}} \times \partial_{\boldsymbol{d}^{\mathrm{eq}}} \bar{\Phi}' = \partial_{\boldsymbol{d}^{\mathrm{in}}} \times \partial_{\boldsymbol{d}^{\mathrm{eq}}} \Phi'$$

on Z', and the claim follows. □

References

1. J. Aires, M. Lima, L. Barroso, P. Lino, M. Pereira, and R. Kelman. The role of demand elasticity in competitive hydrothermal systems. Technical report, Power Systems Research Inc. (PSRI), 2002.
2. R.B. Ash. *Real Analysis and Probability.* Probability and Mathematical Statistics. Academic Press, Berlin, 1972.
3. H. Attouch and R.J.-B. Wets. Approximation and convergence in nonlinear optimization. In O. Mangasarian, R. Meyer, and S.M. Robinson, editors, *Nonlinear Programming 4*, pages 367–394. Academic Press, 1981.
4. L. Bacaud, C. Lemaréchal, A Renaud, and C. Sagastizábal. Bundle methods in stochastic optimal power management: A disaggregated approach using preconditioners. *Computational Optimization and Applications*, 20:227–244, 2001.
5. L. Barroso, M. Fampa, R. Kelman, M. Pereira, and P. Lino. Market power issues in bid-based hydrothermal dispatch. *Annals of Operations Research*, 117:247–270, 2002.
6. M.S. Bazaraa, H.D. Sherali, and C.M. Shetty. *Nonlinear Programming, Theory and Algorithms.* John Wiley & Sons, New York, 2nd edition, 1993.
7. R. Bellman. *Adaptive Control Processes: A Guided Tour.* Princeton University Press, 1961.
8. J.R. Birge. The value of the stochastic solution in stochastic linear programs with fixed recourse. *Mathematical Programming*, 24:314–325, 1982.
9. J.R. Birge and J.H. Dulá. Bounding separable recourse functions with limited distribution information. *Annals of Operations Research*, 30:277–298, 1991.
10. J.R. Birge and F. Louveaux. *Introduction to Stochastic Programming.* Springer-Verlag, New York, 1997.
11. J.R. Birge and R.J.-B. Wets. Designing approximation schemes for stochastic optimization problems, in particular for stochastic programs with recourse. *Mathematical Programming Study*, 27:54–102, 1986.
12. J.R. Birge and R.J.-B. Wets. Computing bounds for stochastic programming problems by means of a generalized moment problem. *Mathematics of Operations Research*, 12:149–162, 1987.
13. K.C. Border. *Fixed Point Theorems with Applications to Economics and Game Theory.* Cambridge University Press, Cambridge, UK, 1985.

14. C.C. Carøe, M.P. Nowak, W. Römisch, and R. Schultz. Power scheduling in a hydro-thermal system under uncertainty. In *Proceedings 13th Power Systems Computation Conference (Trondheim, Norway)*, volume 2, pages 1086–1092, 1999.
15. C.C. Carøe and R. Schultz. Dual decomposition in stochastic integer programming. *Operations Research Letters*, 24:37–45, 1999.
16. CPLEX Optimization Inc. Using the CPLEX callable library. CPLEX Optimization, http://www.ilog.com, 1987–2003.
17. G.B. Danzig. Linear programming under uncertainty. *Management Science*, 1:197–206, 1955.
18. G.B. Danzig. *Linear Programming and Extensions*. Princeton University Press, Princeton, NJ, 1963.
19. S. Deng, B. Johnson, and A. Sogomonian. Exotic electricity options and the valuation of electricity generation and transmission assets. *Decision Support Systems*, 30:383–392, 2001.
20. D. Dentcheva, R. Gollmer, A. Möller, W. Römisch, and R. Schultz. Solving the unit commitment problem in power generation by primal and dual methods. In M. Brøns, M.P. Bendsøe, and M.P. Sørensen, editors, *Progress in Industrial Mathematics at ECMI 96*, pages 332–339. Teubner, 1997.
21. D. Dentcheva and W. Römisch. Optimal power generation under uncertainty via stochastic programming. In K. Marti and P. Kall, editors, *Stochastic Programming Methods and Technical Applications*, volume 458 of *Lecture Notes in Economics and Mathematical Systems*, pages 22–56. Springer-Verlag, 1998.
22. D. Dentcheva and W. Römisch. Duality gaps in nonconvex stochastic optimization. The Stochastic Programming E-Print Series (SPEPS), July 2002.
23. S.P. Dokov and D.P. Morton. Second order lower bounds on the expectation of a convex function. The Stochastic Programming E-Print Series (SPEPS), February 2001.
24. S.P. Dokov and D.P. Morton. Higher-order upper bounds on the expectation of a convex function. The Stochastic Programming E-Print Series (SPEPS), January 2002.
25. J. Dupačová. Minimax stochastic programs with nonconvex nonseparable penalty functions. In A. Prékopa, editor, *Progress in Operations Research*, volume 1 of *Colloquia Mathematica Societatis János Bolyai*, pages 303–316. North-Holland Publishing, Amsterdam, 1976.
26. J. Dupačová. The minimax approach to stochastic programming and an illustrative application. *Stochastics*, 20:73–88, 1987.
27. J. Dupačová. Applications of stochastic programming under incomplete information. *Journal of Computational and Applied Mathematics*, 56:113–125, 1994.
28. J. Dupačová. Stochastic programming: Minimax approach. In C.A. Floudas and P.M. Pardalos, editors, *Encyclopedia of Optimization*, volume 5, pages 327–330. Kluwer, 2001.
29. J. Dupačová, N. Gröwe-Kuska, and W. Römisch. Scenario reduction in stochastic programming: an approach using probability metrics. *Mathematical Programming, Series A*, 95:493–511, 2003.
30. J. Dupačová (as Žáčková). On minimax solutions of stochastic linear programming problems. *Časopis pro Pěstování Matematiky*, 91:423–429, 1966.

31. N.C.P. Edirisinghe. New second-order bounds on the expectation of saddle functions with applications to stochastic linear programming. *Operations Research*, 44:909–922, 1996.
32. N.C.P. Edirisinghe and G.-M. You. Second-order senario approximation and refinement in optimization under uncertainty. *Annals of Operations Research*, 64:143–178, 1996.
33. N.C.P. Edirisinghe and W.T. Ziemba. Bounding the expectation of a saddle function with application to stochastic programming. *Mathematics of Operations Research*, 19:314–340, 1994.
34. N.C.P. Edirisinghe and W.T. Ziemba. Bounds for two-stage stochastic programs with fixed recourse. *Mathematics of Operations Research*, 19:292–313, 1994.
35. H.P. Edmundson. Bounds on the expectation of a convex function of a random variable. Technical report, The Rand Corporation Paper 982, Santa Monica, California, 1956.
36. L.C. Evans and R.F. Gariepy. *Measure Theory and Fine Properties of Functions.* CRC Press, Boca Raton, Florida, 1992.
37. S. Feltenmark and K.C. Kiwiel. Dual applications of proximal bundle methods, including Lagrangian relaxation of nonconvex problems. *SIAM Journal on Control and Optimization*, 10:697–721, 2000.
38. S.-E. Fleten, S. W. Wallace, and W. T. Ziemba. Portfolio management in a deregulated hydropower based electricity Market. In E. Broch, D.K. Lysne, N. Flatabø, and E. Helland-Hansen, editors, *Hydropower 97*, Proceedings of the 3rd international conference on hydropower, Trondheim, Norway, June 30 – July 2, 1997.
39. S.-E. Fleten, S. W. Wallace, and W. T. Ziemba. Hedging electricity portfolios via stochastic programming. In C. Greengard and A. Ruszczynski, editors, *Decision Making Under Uncertainty: Energy and Power*, volume 128 of *IMA Volumes on Mathematics and its Applications*, pages 71–93. Springer-Verlag, 2002.
40. C.A. Floudas. *Deterministic Global Optimization.* Kluwer Academic Publishers, Dordrecht, 2000.
41. K. Frauendorfer. Solving SLP recourse problems with arbitrary multivariate distributions — the dependent case. *Mathematics of Operations Research*, 13:377–394, 1988.
42. K. Frauendorfer. *Stochastic two-stage programming*, volume 392 of *Lecture Notes in Economics and Mathematical Systems.* Springer-Verlag, Berlin, 1992.
43. K. Frauendorfer. Multistage stochastic programming: Error analysis for the convex case. *Z. Oper. Res.*, 39(1):93–122, 1994.
44. K. Frauendorfer. Barycentric scenario trees in convex multistage stochastic programming. *Mathematical Programming*, 75(2):277–294, 1996.
45. K. Frauendorfer and P. Kall. A solution method for SLP recourse problems with arbitrary multivariate distributions — the independent case. *Problems in Control and Information Theory*, 17:177–205, 1988.
46. K. Frauendorfer and M. Schürle. Multistage stochastic programming: Barycentric approximation. In C.A. Floudas and P.M. Pardalos, editors, *Encyclopedia of Optimization*, volume 3, pages 577–580. Kluwer Academic Publishers, 2001.
47. H. Gassmann and W.T. Ziemba. A tight upper bound for the expectation of a convex function of a multivariate random variable. *Mathematical Programming Study*, 27:39–53, 1986.

48. R.C. Grinold. Time horizons in energy planning models. In W.T. Ziemba and S.L. Schwarz, editors, *Energy Policy Modeling: United States and Canadian Experiences*, volume II, pages 216–237. Martinus Nijhoff, Boston, 1980.
49. N. Gröwe-Kuska, K.C. Kiwiel, M.P. Nowak, W. Römisch, and I. Wegner. Power management in a hydro-thermal system under uncertainty by Lagrangian relaxation. The Stochastic Programming E-Print Series (SPEPS), January 2000.
50. N. Gröwe-Kuska and W. Römisch. Stochastic unit commitment in hydrothermal power production planning. To appear in "Applications of Stochastic Programming (S.W. Wallace, W.T. Ziemba eds.), MPS-SIAM Series in Optimization".
51. J. Güssow. *Power Systems Operation and Trading in Competitive Energy Markets*. PhD thesis, IfU-HSG, Universität St. Gallen, 2001.
52. P. Hartman. On functions representable as a difference of convex functions. *Pacific Journal of Mathematics*, 9:707–713, 1959.
53. J.-B. Hiriart-Urruty. Generalized differentiability, duality and optimization for problems dealing with differences of convex functions. In *Convexity and Duality in Optimization (Groningen, 1984)*, volume 256 of *Lecture Notes in Economics and Mathematical Systems*, pages 37–70. Springer-Verlag, 1985.
54. J.-B. Hiriart-Urruty and C. Lemaréchal. *Convex Analysis and Minimization Algorithms II*. Springer-Verlag, Berlin, 1993.
55. R. Horst and H. Tuy. *Global Optimization: Deterministic Approaches*. Springer Verlag, Berlin, 3rd edition, 1996.
56. C.C. Huang, W.T. Ziemba, and A. Ben-Tal. Bounds on the expectation of a convex function of a random variable: With applications to stochastic programming. *Operations Research*, 25:315–325, 1977.
57. J.L. Jensen. Sur les fonctions convexes et les inégalités entre les valeurs moyennes. *Acta Mathematica*, 30:157–193, 1906.
58. P. Kall. Stochastic programming with recourse: upper bounds and moment problems — a review. In J. Guddat, B. Bank, H. Hollatz, P. Kall, D. Klatte, B. Kummer, K. Lommatzch, K. Tammer, M. Vlach, and K. Zimmermann, editors, *Advances in Mathematical Optimization*, pages 86–103. Akademie-Verlag, 1988.
59. P. Kall. An upper bound for SLP using first and total second moments. *Annals of Operations Research*, 30:267–276, 1991.
60. P. Kall, A. Ruszczynski, and K. Frauendorfer. Approximation techniques in stochastic programming. In Y. Ermoliev and R.J.B. Wets, editors, *Numerical Techniques for Stochastic Optimization*, pages 33–64. Springer-Verlag, 1988.
61. P. Kall and S.W. Wallace. *Stochastic Programming*. John Wiley & Sons, Chichester, 1994.
62. M. Kaut and S.W. Wallace. Evaluation of scenario-generation methods for stochastic programming. The Stochastic Programming E-Print Series (SPEPS), May 2003.
63. K.C. Kiwiel. Proximity control in bundle methods for convex nondifferentiable minimization. *Mathematical Programming*, 46:105–122, 1990.
64. W.K. Klein Haneveld. *Duality in Stochastic Linear and Dynamic Programming*, volume 274 of *Lecture Notes in Economics and Mathematical Systems*. Springer-Verlag, Berlin, 1985.
65. M. Leuzinger. *Einsatzplanung hydraulischer Kraftwerke unter stochastischen Bedingungen*. PhD thesis, ETH, Zürich, 1998.

66. M. Loève. *Probability Theory.* Van Nostrand, Princeton, N.J., 3rd edition, 1963.
67. D.T. Luc. *Theory of Vector Optimization*, volume 319 of *Lecture Notes in Economics and Mathematical Systems.* Springer-Verlag, Berlin, 1989.
68. G. Lulli and S. Sen. A branch-and-price algorithm for multistage stochastic integer programs with applications to stochastic lot sizing problems. To appear in "Management Science".
69. A. Madansky. Bounds on the expectation of a convex function of a multivariate random variable. *Annals of Mathematical Statistics*, 30:743–746, 1959.
70. A. Madansky. Inequalities for stochastic linear programming problems. *Management Science*, 6:197–204, 1960.
71. A. Mas-Colell, M.D. Whinston, and J.R. Green. *Microeconomic Theory.* Oxford University Press, New York, Oxford, 1995.
72. Mathematica 4.1. Wolfram Research Inc., http://www.wolfram.com, 2000.
73. A. Möller and W. Römisch. A dual method for the unit commitment problem. Technical report, Humboldt-Universität Berlin, Institut für Mathematik, Preprint Nr. 95-1, 1995.
74. D.P. Morton and R.K. Wood. Restricted-recourse bounds for stochastic linear programming. *Operations Research*, 47:943–956, 1999.
75. M.P. Nowak and W. Römisch. Optimal power dispatch via multistage stochastic programming. In M. Brøns, M.P. Bendsøe, and M.P. Sørensen, editors, *Progress in Industrial Mathematics at ECMI 96*, pages 324–331. Teubner, 1997.
76. M.P. Nowak and W. Römisch. Stochastic Lagrangian relaxation applied to power scheduling in a hydro-thermal system under uncertainty. *Annals of Operations Research*, 100:251–272, 2000.
77. M.P. Nowak, R. Schultz, and M. Westphalen. Optimization of simultaneous power production and trading by stochastic integer programming. The Stochastic Programming E-Print Series (SPEPS), 2002.
78. R. Nürnberg and W. Römisch. A two-stage planning model for power scheduling in a hydro-thermal system under uncertainty. *Optimization and Engineering*, 3:355–378, 2002.
79. G. Ostermaier. *Electric Power System Scheduling by Multistage Stochastic Programming.* PhD thesis, IfU-HSG, Universität St. Gallen, 2001.
80. P.M. Pardalos, H. Edwin Romeijn, and H. Tuy. Recent development and trends in global optimization. *Journal of Computational and Applied Mathematics*, 124:209–228, 2000.
81. M. Pereira, N. Campodónico, and R. Kelman. Long-term hydro scheduling based on stochastic models. In *Proceedings of EPSOM '98, 23–25 Sept. 98, ETH Zürich, Switzerland*, 1998.
82. M. Pereira, N. Campodónico, and R. Kelman. Application of stochastic dual DP and extensions to hydrothermal scheduling. Technical report, Power Systems Research Inc. (PSRI), April 1999. PSRI Technical Report 012/99, Version 2.0.
83. M. Pereira and L. Pinto. Multistage stochastic optimization applied to energy planning. *Mathematical Programming*, 52:359–375, 1991.
84. G.C. Pflug. *Optimization of Stochastic Models.* Kluwer Academic Publishers, Hingham, MA, 1996.
85. G.C. Pflug. Scenario tree generation for multiperiod financial optimization by optimal discretization. *Mathematical Programming, Ser. B*, 89:251–271, 2001.

86. A.B. Philpott, M. Craddock, and H. Waterer. Hydro-electric unit commitment subject to uncertain demand. *European Journal of Operational Research*, 125:410–424, 2000.
87. S.T. Rachev and W. Römisch. Quantitative stability in stochastic programming: the method of probability metrics. *Mathematics of Operations Research*, 27:792–818, 2002.
88. H. Raiffa and R. Schlaifer. *Applied Statistical Decision Theory*. Harvard University, Boston, MA, 1961.
89. R.T. Rockafellar. *Convex Analysis*. Princeton University Press, Princeton, N.J., 1970.
90. R.T. Rockafellar. *Conjugate Duality and Optimization*. SIAM, Philadelphia, 1974.
91. R.T. Rockafellar. Integral functionals, normal integrands and measurable selections. In *Nonlinear Operators and the Calculus of Variations*, volume 543 of *Lecture Notes in Mathematics*, pages 157–207. Springer-Verlag, 1976.
92. R.T. Rockafellar and R.J.-B. Wets. Nonanticipativity and L^1-martingales in stochastic optimization problems. In *Stoch. Syst.: Model., Identif., Optim. II; Math. Program. Study 6*, pages 170–187, 1976.
93. R.T. Rockafellar and R.J.-B. Wets. Stochastic convex programming: Basic duality. *Pacific Journal of Mathematics*, 62:173–195, 1976.
94. R.T. Rockafellar and R.J.-B. Wets. Stochastic convex programming: Relatively complete recourse and induced feasibility. *SIAM Journal on Control and Optimization*, 14:574–589, 1976.
95. R.T. Rockafellar and R.J.-B. Wets. Stochastic convex programming: Singular multipliers and extended duality. *Pacific Journal of Mathematics*, 62:507–522, 1976.
96. R.T. Rockafellar and R.J.-B. Wets. The optimal recourse problem in discrete time: L^1-multipliers for inequality constraints. *SIAM Journal on Control and Optimization*, 16:16–36, 1978.
97. R.T. Rockafellar and R.J.-B. Wets. *Variational Analysis*, volume 317 of *A Series of Comprehensive Studies in Mathematics*. Springer-Verlag, New York, 1998.
98. W. Römisch and R. Schultz. Decomposition of a multi-stage stochastic program for power dispatch. *Zeitschrift für Angewandte Mathematik und Mechanik*, 76(Suppl. 3):29–32, 1996.
99. W. Römisch and R. Schultz. Multistage integer programs: an introduction. In M Grötschel, S.O. Krumke, and J. Rambau, editors, *Online optimization of large scale systems*, pages 581–600. Springer-Verlag, 2001.
100. A. Shapiro. On functions representable as a difference of two convex functions in inequality constrained optimization. Research report, University of South Africa, 1983.
101. J.R. Stedinger. Stochastic multi-reservoir hydroelectric scheduling. In *Proceedings 28. internationales Wasserbau-Symposium, Wasserwirtschaftliche Systeme — Konzepte, Konflikte, Kompromisse*, pages 89–117, Institut für Wasserbau und Wasserwirtschaft, RWTH Aachen, Aachen, Germany, 1998.
102. S. Takriti and J.R. Birge. Lagrangian solution techniques and bounds for loosely coupled mixed-integer stochastic programs. *Operations Research*, 48:91–98, 2000.
103. S. Takriti, J.R. Birge, and E. Long. A stochastic model for the unit commitment problem. *IEEE Transactions on Power Systems*, 11:1497–1508, 1996.

104. S. Takriti, B. Krasenbrink, and L.S.-Y. Wu. Incorporating fuel constraints and electricity spot prices into the stochastic unit commitment problem. *Operations Research*, 48:268–280, 2000.
105. H. Tong. *Nonlinear Time Series — a Dynamical System Approach.* Oxford University Press, Oxford, 1990.
106. S. W. Wallace and S.-E. Fleten. Stochastic programming in energy. In A. Ruszczynski and A. Shapiro, editors, *Stochastic programming*, volume 10 of *Handbooks in Operations Research and Management Science*, pages 637–677. North-Holland, 2003.
107. R.J.-B. Wets. Induced constraints for stochastic optimization problems. In A. Balakrishnan, editor, *Techniques of Optimization*, pages 433–443. Academic Press, 1972.
108. R.J.-B. Wets. Stochastic programming. In G.L. Nemhauser, editor, *Handbooks in OR & MS*, volume 1, pages 573–629. Elsevier Science Publishers B.V., 1989.
109. R.J.-B. Wets. Lipschitz continuity of inf-projections. Technical report, University of California, Davis, 1999.
110. S. Willard. *General Topology.* Addison-Wesley, Reading, MA, 1970.
111. P. Wirth. *Prognosemodelle in der mehrstufigen stochastischen Optimierung.* PhD thesis, IfU-HSG, Universität St. Gallen, 2004.
112. S. Yankowitz. Dynamic programming applications in water resources. *Water Resources Research*, 18:673–696, 1982.
113. W.-W.G. Yeh. Reservoir management and operation models. *Water Resources Research*, 21:1797–1818, 1985.
114. K. Yosida. *Functional Analysis.* Springer-Verlag, Berlin, 6th edition, 1980.
115. W.I. Zangwill. Nonlinear programming via penalty functions. *Management Science*, 13:344–358, 1967.

List of Figures

2.1 Example of a stochastic process 8
2.2 Visualization of the induced filtration 9
2.3 The feasible set mapping X_t 13

3.1 The augmented probability space 31
3.2 Example of a biconcave function 41
3.3 Continuation of a saddle function 44
3.4 Visualization of the generalized feasible sets 48

4.1 The Jensen and Edmundson-Madansky inequalities 58
4.2 Bilinear affine approximations of a saddle function 64
4.3 Bounding set for the optimal first stage decisions 79

5.1 Regularization of a biconcave profit function 85

6.1 Refinement strategy I 129
6.2 Convergence of bounds due to refinements 130
6.3 Bounds in the presence of elastic prices 132
6.4 Refinement strategy II 133
6.5 Adjusted bounds in the presence of lognormal prices 134
6.6 Refinement strategy III 135
6.7 Adjusted bounds in the presence of lognormal inflows 137
6.8 Adjusted bounds in the presence of risk aversion 138

7.1 Interdependence of regularity conditions 144

B.1 Boundedness of the value function 156

List of Tables

6.1 Input parameters . . . 117
6.2 System parameters . . . 127
6.3 Distributional parameters for the reference problem . . . 127
6.4 Results of the reference problem . . . 130
6.5 Piecewise linearization of the inverse demand function . . . 131
6.6 Results in the presence of elastic prices . . . 131
6.7 Distributional parameters in the presence of lognormal prices . . . 132
6.8 Results in the presence of lognormal prices . . . 134
6.9 Distributional parameters in the presence of lognormal inflows . . . 135
6.10 Results in the presence of lognormal inflows . . . 136
6.11 Piecewise linearization of the utility function . . . 138
6.12 Results in the presence of risk aversion . . . 139

Index

adjusted recourse problem, 88, 95, 101
anticipative policy, *see* policy
Attouch, 81
augmented probability space, 30
auxiliary recourse problem, 73
auxiliary stochastic program, 73, 74

barycentric
  coordinates, 54
  probability measures, 71
  transition probabilities, 71
  weights, 54
    history-dependent, 69
  weights (generalized), 55
    history-dependent, 70
barycentric approximation
  lower, 62
  upper, 63
barycentric approximation scheme, 5, 53
biconcave function, 40
biconjugate function, 148
biconvex function, 46
Birge, 3, 6, 27, 46, 65
Border, 16
bounding measure, 53
bounding set, 78, 111
bounds on the recourse functions, 74–77, 88–89, 99–100, 102, 110–111
branch and bound algorithm, 114
branching factor, 53

capacity constraints, 116
Carathéodory map, 17
Carøe, 115
closed function, 148
closure of a function, 147
complementary slackness, 56
component coupling constraints, 115
concave conjugate function, 148
concave function, 35
  on nonconvex domain, 36
  vector-valued, 36
conditional correction terms
  for nonconvex constraints, 95
  for nonconvex objective, 87
  for nonconvex objective *and* constraints, 101
conditional expectation
  regular, 9
    w.r.t. barycentric measures, 71
conditional probability
  regular, 9
    w.r.t. barycentric measures, 71
confidence ellipse, 117
conjugate duality
  see duality, 147
conjugate function, 148
constraint function, 11
  defined on the augmented probability space, 30
constraint multifunction, 12
  non-anticipative, 12
constraint qualification, 33
  Slater, 33–34
convex function, 35
  on nonconvex domain, 35

vector-valued, 35
convex hull, 148
correction terms
conditional, *see* conditional correction term
for linear stochastic programs, 110
for nonconvex constraints, 94
for nonconvex objective, 84
for nonconvex objective *and* constraints, 101
cross-simplex, 54
curse of dimensionality, 53

d.c. function, 105
Danzig, 102
decision history, 10
decision rule, 10
anticipative, 10
non-anticipative, 10
Deng, 120
Dentcheva, 115
Dokov, 3
dual problem
conjugate duality, 150
Lagrangian duality
with equality constraints, 161
without equality constraints, 157
duality
conjugate, 147
Lagrangian, 155
strong, 152
weak, 152
duality gap, 152
Dulá, 3
Dupačová, 3, 56
dynamic constraints, 114
dynamic version, 32

Edirisinghe, 3, 144
Edmundson, 3
Edmundson-Madansky inequality, 59
effective profit function, 14
defined on the augmented probability space, 31
end effects, 127
energy balance equation, 116
energy conversion factor, 118
epi-convergence, 81
event, 7
EVPI, *see* expected value of perfect information
exact penalty functions, 164
expectation functional, 15, 32
expected value of perfect information, 7, 26
extended arithmetic, 36

feasible set mapping, 12, 32
nested, 12, 32
filtration, 7
induced, 8
Fleten, 115
Frauendorfer, 3, 5, 29, 53, 54, 66
Fubini's theorem (generalized), 20

Güssow, 115
Gassmann, 3
generalized barycenters, 61, 63
history-dependent, 70
generalized barycentric weights, *see* barycentric weights
generalized feasible set, 13
defined on the augmented probability space, 32
generalized Fubini theorem, 20
Groewe-Kuska, 115

Hartman, 106
here-and-now, 25
Huang, 65
hypograph, 147

induced constraints, 12
induced filtration, *see* filtration
inverse demand elasticity, 119
inverse demand function, 119

Jensen, 2, 3
Jensen's inequality, 57

Kall, 3, 65
Klein Haneveld, 22

Lagrange multiplier, 161
Lagrangian function, 150, 157, 160
Lagrangian relaxation, 114
level set, 13
linear stochastic program, *see* stochastic program

Louveaux, 27, 46

Madansky, 3, 26
MILP, *see* mixed integer linear program
mixed integer linear program, 114
moment problem, 53
Morton, 3
multifunction, *see* set-valued mapping
multistage stochastic program, *see* stochastic program

natural domain
  of the expectation functional, 15, 32
  of the recourse function, 15, 32
nested feasible set mapping, *see* feasible set mapping
non-anticipative constraint multifunction, *see* constraint multifunction
non-anticipative policy, *see* policy
non-anticipativity constraints, 114
normal form of a linear stochastic program, 104
normal integrand, 12, 19
normality, 152

objective function coefficients, 103
observation, 8
Ostermaier, 115
outcome, 7
outcome history, 10, 30

partition
  of a conditional probability, 68
  of a probability measure, 64
  of a set, 64
penalty formulation, 163–164
Pereira, 114
Pflug, 4, 78
Pinto, 114
policy, 10
  anticipative, 10
  non-anticipative, 10
primal problem
  conjugate duality, 150
  Lagrangian duality
    with equality constraints, 161
    without equality constraints, 157
profit function, 14
  defined on the augmented probability space, 30
proper concave function, 147
proximal bundle method, 114
pseudo-probabilities, 60, 63
  history-dependent, 70

Römisch, 4, 114, 115
Rachev, 4
Raiffa, 26
random variable, 8
random vector, 8
recourse function, 15, 32
recourse matrix, 103
refinement, 65, 71
refinement parameter, 64, 68
regular conditional expectation, *see* conditional expectation
regular cross-simplex, *see* cross-simplex
regular refinement strategy, 72
regularity conditions, 17, 22, 24, 37, 45, 46, 85, 91, 100, 107
regularizable
  constraint function, 89
  profit function, 84
revenue balance equation, 125
rhs vector, *see* right hand side vector
right hand side vector, 103
risk aversion, 123
Rockafellar, 12, 16, 40, 47, 147

sample space, 7
scenario generation, 1, 51
scenario problem, 25
scenario tree, 52
  node, 52
  path probability, 52
Schlaifer, 26
Schultz, 114, 115
SDDP, *see* stochastic dual dynamic programming
SDP, *see* stochastic dynamic programming
set-valued mapping
  bounded, 16
  continuous, 16
  lower semicontinuous, 16
  upper semicontinuous, 16
shadow price, 122
Shapiro, 106
σ-algebra of events, 7

Slater point, 34
Slater's constraint qualification, *see* constraint qualification
stability, 152
state space, 8
static version, 14, 31
stochastic dual dynamic programming, 114
stochastic dynamic programming, 113
stochastic process, 8
  adapted, 8
  block-diagonal autoregressive, 36
  deterministic, 8
  nonlinear autoregressive, 16
  previsible, 8
stochastic program, 1
  convex, 29
  dynamic version, 14, 32
  linear, 102
  non-linear, 14
  static version, 14, 31
strategy, 10
  anticipative, 10
  non-anticipative, 10
subdifferentiability, 47–49
subgradient method, 114
sup-projection, 40, 169–174

technology matrix, 103
transition probability, 52, 68
truncation, 116

upper semicontinuity, 147
usc hull, 147
utility function, 123

value of the stochastic solution, 7, 27
VSS, *see* value of the stochastic solution

wait-and-see, 26
Wallace, 115
weak convergence, 66
Wets, 3, 6, 12, 16, 43, 65, 81, 173

Ziemba, 3, 115

Lecture Notes in Economics and Mathematical Systems

For information about Vols. 1–454
please contact your bookseller or Springer-Verlag

Vol. 455: R. Caballero, F. Ruiz, R. E. Steuer (Eds.), Advances in Multiple Objective and Goal Programming. VIII, 391 pages. 1997.

Vol. 456: R. Conte, R. Hegselmann, P. Terna (Eds.), Simulating Social Phenomena. VIII, 536 pages. 1997.

Vol. 457: C. Hsu, Volume and the Nonlinear Dynamics of Stock Returns. VIII, 133 pages. 1998.

Vol. 458: K. Marti, P. Kall (Eds.), Stochastic Programming Methods and Technical Applications. X, 437 pages. 1998.

Vol. 459: H. K. Ryu, D. J. Slottje, Measuring Trends in U.S. Income Inequality. XI, 195 pages. 1998.

Vol. 460: B. Fleischmann, J. A. E. E. van Nunen, M. G. Speranza, P. Stähly, Advances in Distribution Logistic. XI, 535 pages. 1998.

Vol. 461: U. Schmidt, Axiomatic Utility Theory under Risk. XV, 201 pages. 1998.

Vol. 462: L. von Auer, Dynamic Preferences, Choice Mechanisms, and Welfare. XII, 226 pages. 1998.

Vol. 463: G. Abraham-Frois (Ed.), Non-Linear Dynamics and Endogenous Cycles. VI, 204 pages. 1998.

Vol. 464: A. Aulin, The Impact of Science on Economic Growth and its Cycles. IX, 204 pages. 1998.

Vol. 465: T. J. Stewart, R. C. van den Honert (Eds.), Trends in Multicriteria Decision Making. X, 448 pages. 1998.

Vol. 466: A. Sadrieh, The Alternating Double Auction Market. VII, 350 pages. 1998.

Vol. 467: H. Hennig-Schmidt, Bargaining in a Video Experiment. Determinants of Boundedly Rational Behavior. XII, 221 pages. 1999.

Vol. 468: A. Ziegler, A Game Theory Analysis of Options. XIV, 145 pages. 1999.

Vol. 469: M. P. Vogel, Environmental Kuznets Curves. XIII, 197 pages. 1999.

Vol. 470: M. Ammann, Pricing Derivative Credit Risk. XII, 228 pages. 1999.

Vol. 471: N. H. M. Wilson (Ed.), Computer-Aided Transit Scheduling. XI, 444 pages. 1999.

Vol. 472: J.-R. Tyran, Money Illusion and Strategic Complementarity as Causes of Monetary Non-Neutrality. X, 228 pages. 1999.

Vol. 473: S. Helber, Performance Analysis of Flow Lines with Non-Linear Flow of Material. IX, 280 pages. 1999.

Vol. 474: U. Schwalbe, The Core of Economies with Asymmetric Information. IX, 141 pages. 1999.

Vol. 475: L. Kaas, Dynamic Macroeconomics with Imperfect Competition. XI, 155 pages. 1999.

Vol. 476: R. Demel, Fiscal Policy, Public Debt and the Term Structure of Interest Rates. X, 279 pages. 1999.

Vol. 477: M. Théra, R. Tichatschke (Eds.), Ill-posed Variational Problems and Regularization Techniques. VIII, 274 pages. 1999.

Vol. 478: S. Hartmann, Project Scheduling under Limited Resources. XII, 221 pages. 1999.

Vol. 479: L. v. Thadden, Money, Inflation, and Capital Formation. IX, 192 pages. 1999.

Vol. 480: M. Grazia Speranza, P. Stähly (Eds.), New Trends in Distribution Logistics. X, 336 pages. 1999.

Vol. 481: V. H. Nguyen, J. J. Strodiot, P. Tossings (Eds.). Optimation. IX, 498 pages. 2000.

Vol. 482: W. B. Zhang, A Theory of International Trade. XI, 192 pages. 2000.

Vol. 483: M. Königstein, Equity, Efficiency and Evolutionary Stability in Bargaining Games with Joint Production. XII, 197 pages. 2000.

Vol. 484: D. D. Gatti, M. Gallegati, A. Kirman, Interaction and Market Structure. VI, 298 pages. 2000.

Vol. 485: A. Garnaev, Search Games and Other Applications of Game Theory. VIII, 145 pages. 2000.

Vol. 486: M. Neugart, Nonlinear Labor Market Dynamics. X, 175 pages. 2000.

Vol. 487: Y. Y. Haimes, R. E. Steuer (Eds.), Research and Practice in Multiple Criteria Decision Making. XVII, 553 pages. 2000.

Vol. 488: B. Schmolck, Ommitted Variable Tests and Dynamic Specification. X, 144 pages. 2000.

Vol. 489: T. Steger, Transitional Dynamics and Economic Growth in Developing Countries. VIII, 151 pages. 2000.

Vol. 490: S. Minner, Strategic Safety Stocks in Supply Chains. XI, 214 pages. 2000.

Vol. 491: M. Ehrgott, Multicriteria Optimization. VIII, 242 pages. 2000.

Vol. 492: T. Phan Huy, Constraint Propagation in Flexible Manufacturing. IX, 258 pages. 2000.

Vol. 493: J. Zhu, Modular Pricing of Options. X, 170 pages. 2000.

Vol. 494: D. Franzen, Design of Master Agreements for OTC Derivatives. VIII, 175 pages. 2001.

Vol. 495: I. Konnov, Combined Relaxation Methods for Variational Inequalities. XI, 181 pages. 2001.

Vol. 496: P. Weiß, Unemployment in Open Economies. XII, 226 pages. 2001.

Vol. 497: J. Inkmann, Conditional Moment Estimation of Nonlinear Equation Systems. VIII, 214 pages. 2001.

Vol. 498: M. Reutter, A Macroeconomic Model of West German Unemployment. X, 125 pages. 2001.

Vol. 499: A. Casajus, Focal Points in Framed Games. XI, 131 pages. 2001.

Vol. 500: F. Nardini, Technical Progress and Economic Growth. XVII, 191 pages. 2001.

Vol. 501: M. Fleischmann, Quantitative Models for Reverse Logistics. XI, 181 pages. 2001.

Vol. 502: N. Hadjisavvas, J. E. Martínez-Legaz, J.-P. Penot (Eds.), Generalized Convexity and Generalized Monotonicity. IX, 410 pages. 2001.

Vol. 503: A. Kirman, J.-B. Zimmermann (Eds.), Economics with Heterogenous Interacting Agents. VII, 343 pages. 2001.

Vol. 504: P.-Y. Moix (Ed.), The Measurement of Market Risk. XI, 272 pages. 2001.

Vol. 505: S. Voß, J. R. Daduna (Eds.), Computer-Aided Scheduling of Public Transport. XI, 466 pages. 2001.

Vol. 506: B. P. Kellerhals, Financial Pricing Models in Continuous Time and Kalman Filtering. XIV, 247 pages. 2001.

Vol. 507: M. Koksalan, S. Zionts, Multiple Criteria Decision Making in the New Millenium. XII, 481 pages. 2001.

Vol. 508: K. Neumann, C. Schwindt, J. Zimmermann, Project Scheduling with Time Windows and Scarce Resources. XI, 335 pages. 2002.

Vol. 509: D. Hornung, Investment, R&D, and Long-Run Growth. XVI, 194 pages. 2002.

Vol. 510: A. S. Tangian, Constructing and Applying Objective Functions. XII, 582 pages. 2002.

Vol. 511: M. Külpmann, Stock Market Overreaction and Fundamental Valuation. IX, 198 pages. 2002.

Vol. 512: W.-B. Zhang, An Economic Theory of Cities.XI, 220 pages. 2002.

Vol. 513: K. Marti, Stochastic Optimization Techniques. VIII, 364 pages. 2002.

Vol. 514: S. Wang, Y. Xia, Portfolio and Asset Pricing. XII, 200 pages. 2002.

Vol. 515: G. Heisig, Planning Stability in Material Requirements Planning System. XII, 264 pages. 2002.

Vol. 516: B. Schmid, Pricing Credit Linked Financial Instruments. X, 246 pages. 2002.

Vol. 517: H. I. Meinhardt, Cooperative Decision Making in Common Pool Situations. VIII, 205 pages. 2002.

Vol. 518: S. Napel, Bilateral Bargaining. VIII, 188 pages. 2002.

Vol. 519: A. Klose, G. Speranza, L. N. Van Wassenhove (Eds.), Quantitative Approaches to Distribution Logistics and Supply Chain Management. XIII, 421 pages. 2002.

Vol. 520: B. Glaser, Efficiency versus Sustainability in Dynamic Decision Making. IX, 252 pages. 2002.

Vol. 521: R. Cowan, N. Jonard (Eds.), Heterogenous Agents, Interactions and Economic Performance. XIV, 339 pages. 2003.

Vol. 522: C. Neff, Corporate Finance, Innovation, and Strategic Competition. IX, 218 pages. 2003.

Vol. 523: W.-B. Zhang, A Theory of Interregional Dynamics. XI, 231 pages. 2003.

Vol. 524: M. Frölich, Programme Evaluation and Treatment Choise. VIII, 191 pages. 2003.

Vol. 525:S. Spinler, Capacity Reservation for Capital-Intensive Technologies. XVI, 139 pages. 2003.

Vol. 526: C. F. Daganzo, A Theory of Supply Chains. VIII, 123 pages. 2003.

Vol. 527: C. E. Metz, Information Dissemination in Currency Crises. XI, 231 pages. 2003.

Vol. 528: R. Stolletz, Performance Analysis and Optimization of Inbound Call Centers. X, 219 pages. 2003.

Vol. 529: W. Krabs, S. W. Pickl, Analysis, Controllability and Optimization of Time-Discrete Systems and Dynamical Games. XII, 187 pages. 2003.

Vol. 530: R. Wapler, Unemployment, Market Structure and Growth. XXVII, 207 pages. 2003.

Vol. 531: M. Gallegati, A. Kirman, M. Marsili (Eds.), The Complex Dynamics of Economic Interaction. XV, 402 pages, 2004.

Vol. 532: K. Marti, Y. Ermoliev, G. Pflug (Eds.), Dynamic Stochastic Optimization. VIII, 336 pages. 2004.

Vol. 533: G. Dudek, Collaborative Planning in Supply Chains. X, 234 pages. 2004.

Vol. 534: M. Runkel, Environmental and Resource Policy for Consumer Durables. X, 197 pages. 2004.

Vol. 535: X. Gandibleux, M. Sevaux, K. Sörensen, V. T'kindt (Eds.), Metaheuristics for Multiobjective Optimisation. IX, 249 pages. 2004.

Vol. 536: R. Brüggemann, Model Reduction Methods for Vector Autoregressive Processes. X, 218 pages. 2004.

Vol. 537: A. Esser, Pricing in (In)Complete Markets. XI, 122 pages, 2004.

Vol. 538: S. Kokot, The Econometrics of Sequential Trade Models. XI, 193 pages. 2004.

Vol. 539: N. Hautsch, Modelling Irregularly Spaced Financial Data. XII, 291 pages. 2004.

Vol. 540: H. Kraft, Optimal Portfolios with Stochastic Interest Rates and Defaultable Assets. X, 173 pages. 2004.

Vol. 541: G.-Y. Chen, Vector Optimization (planned).

Vol. 542: J. Lingens, Union Wage Bargaining and Economic Growth. XIII, 199 pages. 2004.

Vol. 543: C. Benkert, Default Risk in Bond and Credit Derivatives Markets. IX, 135 pages. 2004.

Vol. 544: B. Fleischmann, A. Klose, Distribution Logistics. X, 284 pages. 2004.

Vol. 545: R. Hafner, Stochastic Implied Volatility. XI, 229 pages. 2004.

Vol. 546: D. Quadt, Lot-Sizing and Scheduling for Flexible Flow Lines. XVIII, 227 pages. 2004.

Vol. 547: M. Wildi, Signal Extraction. XI, 279 pages. 2005.

Vol. 548: D. Kuhn, Generalized Bounds for Convex Multistage Stochastic Programs. XI, 190 pages. 2005.

Printing and Binding: Strauss GmbH, Mörlenbach

LECTURE NOTES IN ECONOMICS AND MATHEMATICAL SYSTEMS

This series reports on new developments in mathematical economics, economic theory, econometrics, operations research, and mathematical systems. The type of material considered for publication includes

1. Research monographs
2. Lectures on a new field or presentations of a new angle in a classical field
3. Seminars on topics of current research
4. Reports of meetings, provided they are
   - of exceptional interest and
   - devoted to a single topic.

Texts which are out of print but still in demand may also be considered if they fall within these categories.

In the case of a research monograph, or of seminar notes, the timeliness of a manuscript may be more important than its form, which may be preliminary or tentative. A subject index should be included.

Manuscripts

Manuscripts should be no less than 150 and preferably no more than 500 pages in length. They should be submitted in camera-ready form according to Springer's specifications. Technical instructions will be sent on request. Manuscripts should be sent directly to Springer Heidelberg.

ISSN 0075-8442

ISBN 3-540-22450-4

9 783540 225409

springeronline.com